CONTEMPORARY MATHEMATICS

Titles in this Series

Volume

1 **Markov random fields and their applications**, Ross Kindermann and J. Laurie Snell

2 **Proceedings of the conference on integration, topology, and geometry in linear spaces**, William H. Graves, Editor

3 **The closed graph and P-closed graph properties in general topology**, T. R. Hamlett and L. L. Herrington

4 **Problems of elastic stability and vibrations**, Vadim Komkov, Editor

5 **Rational constructions of modules for simple Lie algebras**, George B. Seligman

6 **Umbral calculus and Hopf algebras**, Robert Morris, Editor

7 **Complex contour integral representation of cardinal spline functions**, Walter Schempp

8 **Ordered fields and real algebraic geometry**, D. W. Dubois and T. Recio, Editors

9 **Papers in algebra, analysis and statistics**, R. Lidl, Editor

10 **Operator algebras and K-theory**, Ronald G. Douglas and Claude Schochet, Editors

11 **Plane ellipticity and related problems**, Robert P. Gilbert, Editor

12 **Symposium on algebraic topology in honor of José Adem**, Samuel Gitler, Editor

13 **Algebraists' homage: Papers in ring theory and related topics**, S. A. Amitsur, D. J. Saltman and G. B. Seligman, Editors

14 **Lectures on Nielsen fixed point theory**, Boju Jiang

15 **Advanced analytic number theory. Part I: Ramification theoretic methods**, Carlos J. Moreno

16 **Complex representations of GL(2, K) for finite fields K**, Ilya Piatetski-Shapiro

17 **Nonlinear partial differential equations**, Joel A. Smoller, Editor

18 **Fixed points and nonexpansive mappings**, Robert C. Sine, Editor

19 **Proceedings of the Northwestern homotopy theory conference**, Haynes R. Miller and Stewart B. Priddy, Editors

20 **Low dimensional topology**, Samuel J. Lomonaco, Jr., Editor

21 **Topological methods in nonlinear functional analysis**, S. P. Singh, S. Thomeier, and B. Watson, Editors

22 **Factorizations of $b^n \pm 1$, $b = 2, 3, 5, 6, 7, 10, 11, 12$ up to high powers**, John Brillhart, D. H. Lehmer, J. L. Selfridge, Bryant Tuckerman, and S. S. Wagstaff, Jr.

23 **Chapter 9 of Ramanujan's second notebook—Infinite series identities, transformations, and evaluations**, Bruce C. Berndt and Padmini T. Joshi

24 **Central extensions, Galois groups, and ideal class groups of number fields**, A. Fröhlich

25 **Value distribution theory and its applications**, Chung-Chun Yang, Editor

26 **Conference in modern analysis and probability**, Richard Beals, Anatole Beck, Alexandra Bellow and Arshag Hajian, Editors

27 **Microlocal analysis**, M. Salah Baouendi, Richard Beals and Linda Preiss Rothschild, Editors

Titles in this Series

Volume

28 **Fluids and plasmas: geometry and dynamics**, Jerrold E. Marsden, Editor

29 **Automated theorem proving**, W. W. Bledsoe and Donald Loveland, Editors

30 **Mathematical applications of category theory**, J. W. Gray, Editor

31 **Axiomatic set theory**, James E. Baumgartner, Donald A. Martin and Saharon Shelah, Editors

32 **Proceedings of the conference on Banach algebras and several complex variables**, F. Greenleaf and D. Gulick, Editors

33 **Contributions to group theory**, Kenneth I. Appel, John G. Ratcliffe and Paul E. Schupp, Editors

34 **Combinatorics and algebra**, Curtis Greene, Editor

35 **Four-manifold theory**, Cameron Gordon and Robion Kirby, Editors

36 **Group actions on manifolds**, Reinhard Schultz, Editor

37 **Conference on algebraic topology in honor of Peter Hilton**, Renzo Piccinini and Denis Sjerve, Editors

38 **Topics in complex analysis**, Dorothy Brown Shaffer, Editor

39 **Errett Bishop: Reflections on him and his research**, Murray Rosenblatt, Editor

40 **Integral bases for affine Lie algebras and their universal enveloping algebras**, David Mitzman

41 **Particle systems, random media and large deviations**, Richard Durrett, Editor

42 **Classical real analysis**, Daniel Waterman, Editor

43 **Group actions on rings**, Susan Montgomery, Editor

44 **Combinatorial methods in topology and algebraic geometry**, John R. Harper and Richard Mandelbaum, Editors

Classical Real Analysis

CASPER GOFFMAN

CONTEMPORARY MATHEMATICS

Volume 42

Classical Real Analysis

Daniel Waterman, Editor

AMERICAN MATHEMATICAL SOCIETY

Providence · Rhode Island

PROCEEDINGS OF THE SPECIAL SESSION ON CLASSICAL REAL ANALYSIS
794TH MEETING OF THE AMERICAN MATHEMATICAL SOCIETY

HELD IN MADISON, WISCONSIN
APRIL 16–17, 1982
(Several papers contributed in honor of Casper Goffman)

1980 *Mathematics Subject Classification.* Primary 26-06; Secondary 26A24, 26A45.

Library of Congress Cataloging-in-Publication Data

Special Session on Classical Real Analysis (1982: Madison, Wis.)
Classical real analysis.

(Contemporary mathematics, ISSN 0271-4132; v. 42)
"Proceedings of the Special Session on Classical Real Analysis, 794th meeting of the American Mathematical Society, held in Madison, Wis. April 16–17, 1982"– Verso t.p.
Bibliography: p.
1. Functions of real variables—Congresses. I. Waterman, Daniel. II. American Mathematical Society. Meeting. (794th : 1982 : Madison, Wis.) III. Title. IV. Series: Contemporary mathematics (American Mathematical Society) v. 42.
QA331.5.S64 1982 515.8 85-9241
ISBN 0-8218-5045-8 (alk. paper)

TABLE OF CONTENTS

Introduction ix

Cesari Spaces and Sobolev Spaces in Surface Area and Localization for Multiple Fourier Series
C. GOFFMAN 1

An Unusual Descriptive Definition of Integral
B. BONGIORNO and D. PREISS 13

Path Derivatives: A Unified View of Certain Generalized Derivatives
A. M. BRUCKNER, R. J. O'MALLEY and B. S. THOMSON 23

Baire One, Null Functions
A. M. BRUCKNER, J. MAŘÍK, and C. E. WEIL 29

Monotone Approximation on an Interval
R. DARST and R. HUOTARI 43

Two Remarks on the Measure of Product Sets
R. O. DAVIES 45

Monotonicity, Symmetry, and Smoothness
M. J. EVANS and L. LARSON 49

The Structure of Continuous Functions which Satisfy Lusin's Condition (N)
J. FORAN 55

Construction of Absolutely Continuous and Singular Functions That are Nowhere of Monotonic Type
K. GARG 61

A Porosity Characterization of Symmetric Perfect Sets
P. D. HUMKE and B. S. THOMSON 81

A Method for Showing Generalized Derivatives are in Baire Class One
L. LARSON 87

On Generalizations of Exact Peano Derivatives
C.-M. LEE 97

Best Monotone Approximation in $L_\infty[0,1]$
D. LEGG 105

Representation of Lattices and Extension of Measures
F. C. LIU 113

Transformation and Multiplication of Derivatives
J. MAŘÍK 119

A Lusin Type Approximation of Sobolev Functions by Smooth Functions
J. H. MICHAEL and W. P. ZIEMER 135

Some Properties of Fourier Series With Gaps
C. J. NEUGEBAUER 169

An Extension of Thunsdorff's Integral Inequality to a Class of Monotone Functions
T. NISHIURA 175

On Generalized Bounded Variation
L. Di PIAZZA and C. MANISCALCO 179

On the Level Set Structure of a Continuous Function
B. S. THOMSON 187

Some Properties of the Littlewood-Paley g-Function
S. WANG 191

Change-of-Variable Invariant Classes of Functions and Convergence of Fourier Series
D. WATERMAN 203

Schauder Bases for $L^p[0,1]$ Derived From Subsystems of the Schauder System
R. E. ZINK 213

INTRODUCTION

This volume contains most of the papers presented at a Special Session on Classical Real Analysis organized in honor of Casper Goffman at the 794th meeting of the American Mathematical Society held at the University of Wisconsin, Madison, April 16-17, 1982. The paper of Goffman which appears here is based on his one-hour address at that meeting.

Many mathematicians who did not present papers in that session have also contributed papers to this volume. The papers which appear here are offered in admiration and respect for the example Professor Goffman has set us and in gratitude for the friendship, encouragement, and advice he has so generously given.

INTRODUCTION

Contemporary Mathematics
Volume 42, 1985

CESARI SPACES AND SOBOLEV SPACES IN SURFACE AREA AND LOCALIZATION FOR MULTIPLE FOURIER SERIES

Casper Goffman[1]

1. We are interested in n dimensional analogues of three classes of functions of one real variable. These are the set, BV, of functions of bounded variation, the set, CBV, of continuous functions of bounded variation, and the set, AC, of absolutely continuous functions.

As a distribution, $f \in BV$ if and only if its derivative is a finite measure, $f \in CBV$ if and only if this measure is nonatomic, $f \in AC$ if and only if this measure is absolutely continuous with respect to Lebesgue measure.

Arc length may be defined for functions in BV which are continuous at the end points and for which the value at each point lies between the right and left handed limits at that point in the same way as for functions in CBV. Let $f : [a,b] \to R$ be of this sort. For each $(\alpha,\beta) \subset [a,b]$, let $\ell(f;(\alpha,\beta))$ denote the length of the curve given by f on (α,β). This function of open intervals may be extended to a measure $\ell_f(S) = \ell(f;S)$ on the Borel sets $S \subset [a,b]$. We call this the length measure.

The following approximation theorems hold.

a) If $f \in BV$ then, for every $\epsilon > 0$, there is a continuously differentiable g such that $f(x) = g(x)$ except on a set of measure less than ϵ and $|\ell(f;[a,b]-\ell(g,[a,b])| < \epsilon$, [17].

b) $f \in AC$ if and only if, for every $\epsilon > 0$, there is a set E and a continuously differentiable g such that $f(x) = g(x)$ on $[a,b] \setminus E$ and $\ell(f;E) < \epsilon$, $\ell(g;E) < \epsilon$, [17].

c) $f \in CBV$ if and only if, for every $\epsilon > 0$, there is a set E and a g, with continuously turning tangent, such that $f(x) = g(x)$ on $[a,b] \setminus E$ and $\ell(f;E) < \epsilon$, $\ell(g;E) < \epsilon$, [22].

1980 Mathematics Subject Classification. 28A75, 26B15, 42B05, 46E35.

[1]Supported in part by the National Science Foundation.

We also wish to note that $f \in CBV$ if and only if $\ell(f;[a,b])<\infty$ and the Hausdorff 1-dimensional measure of the graph of f is equal to $\ell(f;[a,b])$.

2. The set BVC, [2], [5] is an n-dimensional analogue of the set BV in one dimension. Let $(x_1,\ldots,x_n)$ be a rectangular coordinate system. $f \in BVC$ if, for each $i=1,\ldots,n$, there is an f_i equivalent to f which is of bounded variation in x_i for almost all values of the remaining n-1 variables, and these variation functions are all summable. It is a fact, [26], that a single function equivalent to f has the stated property for all directions.

If the functions f_i for the definition of $f \in BVC$ may be taken to be continuous as well as of bounded variation in x_i for almost all values of the other n-1 variables, then we say that f is linearly continuous. The set of linearly continuous functions is designated by $\mathcal{L}$. Clearly $\mathcal{L} \subset BVC$. Moreover, if the f_i are absolutely continuous in x_i for almost all values of the other variables, then $f \in W_1^1$.

The functions $f \in BVC$ are those whose partial derivatives, in the distribution sense, are finite measures, and the $f \in W_1^1$ are those whose partial derivatives are given by summable functions. A deeper analysis, [18], shows that not all measures can be partial derivatives of functions in BVC. Indeed, all such measures must be absolutely continuous with respect to Hausdorff (n-1)-dimensional measure. Furthermore, partial derivatives of functions in $\mathcal{L}$ have measure zero for all sets of finite Hausdorff (n-1)-dimensional measure.

Moreover, the class $\mathcal{L}$ may be characterized as the set of functions for which the nonparametric surface area is equal to the Hausdorff (n-1)-dimensional measure of the graph, suitably taken. It is in this context that the class $\mathcal{L}$ was introduced, [6]. It consists of those functions in BVC which are approximately continuous everywhere, except on a set of Hausdorff (n-1)-dimensional measure zero, [8]. For these and other reasons, $\mathcal{L}$ seems to be the appropriate analogue in n-space of CBV. Note that for $n=1$, sets of Hausdorff (n-1)-dimensional measure zero are simply the empty sets, and for functions in BV, approximate continuity reduces to continuity.

We also give analogues of the approximation theorems. In order to do this, we must first define nonparametric surface area for functions in BVC and introduce the related area measure. Let

$\mu_1,\ldots,\mu_n$ be the measures which are the partial derivatives of $f \in BVC$ and let λ be Lebesgue measure. The area measure is the variation measure α associated with the vector valued measure $(\lambda,\mu_1,\ldots,\mu_n)$. This measure may also be obtained by associating a surface area of f over every open interval R as the lim inf of the areas of piecewise linear functions converging to f in the L_1 norm, [5]. The resulting function of intervals may be extended to a measure on the Borel sets which agrees with α. If more than one function is involved, we use the notation $\alpha_f, \alpha_g, \ldots$.

We may now state the approximation theorems.

a) For $Q \subset R^n$, $f: Q \to R^1$, $f \in BVC$, and any $\epsilon > 0$, there is a continuously differentiable g such that $f(x) = g(x)$ except on a set of measure less than ϵ and $|\alpha_f(Q) - \alpha_g(Q)| < \epsilon$, [24], [7].

b) $f \in W_1^1$ if and only if, for every $\epsilon > 0$, there is a continuously differentiable g such that if $E = [x: f(x) \neq g(x)]$ then $\alpha_f(E) < \epsilon$ and $\alpha_g(E) < \epsilon$.

c) For $\mathcal{L}$ our knowledge is incomplete.

(i) For $n = 2$, if $f \in \mathcal{L}$, then, for every $\epsilon > 0$, there is a continuous g such that if $E = [x: f(x) \neq g(x)]$, then $\alpha_f(E) < \epsilon$ and $\alpha_g(E) < \epsilon$, [13].

(ii) For $n > 2$, if $f \in \mathcal{L}$, then for every $\epsilon > 0$ there is an approximately continuous g such that if $E = [x: f(x) \neq g(x)]$, then $\alpha_f(E) < \epsilon$ and $\alpha_g(E) < \epsilon$, [8].

It follows from c) that $\mathcal{L}$ is coordinate invariant.

The related result that $f \in \mathcal{L}$ if and only if it is approximately continuous except on a set of Hausdorff n-1 dimensional measure zero is not as deep.

3. This section considers some one dimensional facts needed in the sequel. Let $I = [a,b]$, and let $\{\lambda_n\}$ be an increasing sequence of positive numbers which is unbounded and such that $\Sigma \frac{1}{\lambda_n} = \infty$. We define Λ bounded variation. Let $I_n = [a_n, b_n]$, $n = 1, 2, \ldots$, be nonoverlapping closed intervals in $[a,b]$, and let

$$v(\Lambda, f, \{I_n\}) = \Sigma \frac{1}{\lambda_n} |f(b_n) - f(a_n)| .$$

The Λ variation, $V_\Lambda(f)$, is defined to be the supremum of the $v(\Lambda, f, \{I_n\})$ for all admissible $\{I_n\}$. f is of Λ bounded variation if $V_\Lambda(f)$ is finite, and we say $f \in \Lambda BV$. The special case $\lambda_n = n$ is of particular interest, is designated HBV, and is called harmonic bounded variation.

The main fact, [29], is the following:

a) If $f \in HBV$ then the Fourier series of f converges everywhere. For any $\Lambda BV \not\subset HBV$ there is an $f \in \Lambda BV$ whose Fourier series diverges somewhere.

General classes of variation functions were studied by Salem and L. C. Young and by Garsia and Sawyer. Although the two kinds of extensions discussed by these authors seem to be unrelated, we have the following:

b) HBV includes the classes considered by both sets of authors above, [11], [29].

There is also a connection with another sort of question. A necessary and sufficient condition has been given, [30], for a function f to be such that, for all homeomorphisms g of $[-\pi,\pi]$, the Fourier series of $f \circ g$ converges everywhere. The condition is complicated. HBV is a proper subset of the class of regulated functions satisfying this condition.

4. We turn now to the Lebesgue area of continuous parametric mappings. We are interested in the relation between this area and the value of the integral involving ordinary Jacobians when they exist. Let Q be a closed n-cube, $m \geq n$, and $f: Q \to R^m$ a continuous mapping. Then $f = (f_1, \ldots, f_m)$, where $f_i = f_i(u_1, \ldots, u_n)$, $i = 1, \ldots, m$. The Lebesgue area, $A(f)$, is the lim inf of the areas of piecewise linear mappings converging uniformly to f.

Suppose first that $n = 2$. Let J be the integrand for the classical area formula, using ordinary Jacobians. It is not difficult to prove, [9], that if $E \subset Q$ is a measurable set at each point of which the Jacobians exist then $A(f) \geq \int_E J\, du_1 du_2$.

It is a surprising fact this no longer holds for $n > 2$. The point of departure is a classical example, [1], of a continuous mapping from the 2-cube into 3-space which is a homeomorphism and is such that the Lebesgue area of the surface exceeds the 3-dimensional measure of its graph.

We indicate the construction of the example in [1]. In the closed 3-cube A_0, consider eight pairwise disjoint closed cubes, disjoint with ∂A_0, whose union is designated by A_1. In the interior of each cube of A_1 consider eight pairwise disjoint closed cubes. Designate the union of the 8^2 cubes by A_2. In this way, obtain $A_0 \supset A_1 \supset A_2 \supset \ldots \supset A_k \supset \ldots$. Moreover, choose the sets A_k so that the 3-dimensional Lebesgue measure $\lambda_3(A_k)$ of each exceeds $\frac{1}{2}$.

Let Q_2 be the closed unit square and let p_1 be a piecewise linear mapping of Q_2 into $A_0 \setminus A_1^0$ whose image meets the boundary of each cube of A_1 in a square. Alter p_1 on ∂A_1 to obtain a piecewise linear p_2 of Q_2 into $Q_0 \setminus A_2^0$ whose image meets the boundary of each cube of A_2 in a square. Continue this process to obtain a sequence $\{p_k\}$ of piecewise linear mappings. The $\{p_k\}$ may be chosen so that their areas are as small as we please, say all less than $\frac{1}{4}$. Also, the sequence $\{p_k\}$ converges uniformly to a continuous mapping f. The mappings $\{p_k\}$ may be chosen to be one-one and from their manner of definition it follows that f is one-one. Moreover the image $f(Q_2)$ of Q_2 under f contains $A = \cap A_k$ so that $\lambda_3(f(Q_2)) \geq \frac{1}{2}$. But the Lebesgue area of the surface given by f is $A(f) \leq \liminf A(p_k) \leq \frac{1}{4}$.

Note that the zero dimensional set A of positive Lebesgue 3-dimensional measure does not contribute to the surface area of f. This fact suggests that there could be a continuous mapping f from Q_3 into R^3 for which an appropriate one-dimensional set does not contribute to the surface area. This set should have positive 3-dimensional measure and be the image under f of a set on which the integral of the Jacobian is positive. The possibility of such an example is further suggested by the fact that the Jacobian at a point depends only on the values of the mapping on the lines through the point in the coordinate directions. Indeed, we have constructed such an example in [10]. There is a continuous $f: Q_3 \to R^3$ for which the Jacobian exists on a set E of positive measure and

$$A(f) < \int_E J\, du_1 du_2 du_3 ,$$

where $A(f)$ is the area of the total mapping from Q_3 into R^3.

5. The fact that for $n = 3$ the inequality can go the unaccustomed way motivates the search for conditions under which the area formula holds. We work in a setting which is somewhat more general. For nonparametric surfaces, area is lower semicontinuous for metrices "more general" than the uniform convergence metric, e.g. the L_1-metric and the convergence in measure metric. For parametric mappings, however, for any continuous mapping there is a sequence of mappings converging to it, even everywhere, for which the area converges to zero.

We first consider mappings from 2-space into m-space, $m \geq 2$. We define area for linearly continuous mappings $f = (f_1, \ldots, f_m)$.

Then each f_i, $i = 1,\dots,m$, is continuous in u_1 for almost all u_2 and in u_2 for almost all u_1. We impose a metric on the set of piecewise linear mappings for which the completion is the set of linearly continuous mappings, [14]. Convergence in this metric is called linear convergence and amounts to uniform convergence on almost all lines parallel to the coordinate axes. Area for piecewise linear mappings is lower semicontinuous with respect to this metric and so yields an area for each linearly continuous mapping. For continuous mappings this area agrees with Lebesgue area.

We call a mapping a Sobolev mapping if each $f_i \in W^1_{p_i}$, $i = 1,\dots,m$ and for each $1 \le i < j \le m$, $\frac{1}{p_i} + \frac{1}{p_j} \le 1$. For $n = 2$, the Sobolev mappings have an area since they are linearly continuous. Moreover, for Sobolev mappings, the number

$$I(f) = \int_Q J\, du_1 du_2$$

is defined and is finite. We consider continuously differentiable mollifiers (regularizers). It is a fact, [20], that if f is a Sobolev mapping the mollified functions, f_k, converge linearly to f, $I(f_k) = A(f_k)$, and $I(f) = \lim I(f_k)$. But $A(f) \le \lim A(f_k)$. So $A(f) \le I(f)$. For $n = 2$, the reverse inequality also holds, [14]. For this we need a sharpening, due to Radó, of the classical Stepanov theorem. Stepanov's theorem says that if the approximate partials exist almost everywhere, then the approximate total differential exists almost everywhere. The fact we need is that, for $n = 2$, if the partial derivatives themselves exist almost everywhere, then the regular approximate total differential exists almost everywhere. This means that the set of density one at a point with respect to which the total differential exists is composed of the boundaries of oriented squares with the point as center. For a Sobolev mapping the partial derivatives exist almost everywhere. It follows without too much difficulty that for a Sobolev mapping f from 2-space into m-space, $m \ge 2$,

$$A(f) = \int_Q J\, du_1 du_2 .$$

6. For $n > 2$, the situation is quite complicated. Indeed, the area of piecewise linear mappings is no longer lower semicontinuous with respect to linear convergence. In [14], an example is given of a piecewise linear mapping p, with $A(p) = 1$, and a sequence $\{p_k\}$ converging linearly to p; with $\lim_{k\to\infty} A(p_k) = 0$. On

the other hand, area is lower semicontinuous with respect to (n-1)-convergence. This amounts to uniform convergence on almost all hyperplanes parallel to coordinate hyperplanes. It yields a surface area for all (n-1)-continuous mapping, which agrees with Lebesgue area for all continuous mappings. The (n-1)-continuous mappings are those which are continuous on almost all hyperplanes parallel to the coordinate hyperplanes, [20].

For $n > 2$, the domain of this area functional does not contain all Sobolev mappings since the latter are linearly continuous but not necessarily (n-1)-continuous. We consider Sobolev mappings $f = (f_1, \ldots, f_m)$, of n-space into m-space, $m \geq n$, for which the coordinate functions f_i are in $W^1_{p_i}$, $p_i > n-1$, $i = 1, \ldots, m$, [20]. Since f is a Sobolev mapping, we have $\sum_{j=1}^{m} \frac{1}{p_{ij}} \leq 1$ for every $1 \leq i_1 < \ldots < i_m \leq n$. This implies that $I(f) = \int_Q J$ exists and is finite, and for the same argument as for $n = 2$ we have $I(f) \geq A(f)$. Our condition $p_i > n-1$, $i = 1, \ldots, m$, forces the (n-1)-continuity of f and thus the existence of $A(f)$. It also implies the existence of regular approximate differentials almost everywhere; this means that the set of density one at a point with respect to which the total differential exists is comprised of the boundaries of oriented cubes with the point as center. The existence of the regular approximate differential almost everywhere follows by an application of the Sobolev-Morrey inequality which asserts that if $f: R^n \to R^1$ is in W^1_p, $p > n$, then f has a total differentiable almost everywhere. Applying this fact to hyperplanes when $f \in W^1_p$, $p > n-1$ we obtain, with some work, the existence of the regular approximate differential almost everywhere. It then follows, as for $n = 2$, that

$$A(f) = I(f).$$

7. There are other approaches which seem to be more general. Let $f = (f_1, \ldots, f_m)$ be a Sobolev mapping of $Q \subset R^n$ into R^m, $m \geq n$. For each $x \in Q$ and $t > 0$, let $C_x(t)$ be the oriented cube of center x and side length $2t$ and let $B_x(t)$ be the boundary of $C_x(t)$. The mapping f is regular if, for almost every $x \in Q$, there is a set $S_x \subset R^+$, of positive measure in every neighborhood of 0, such that f restricted to $B_x(t)$ is continuous for every $t \in S_x$ and

$$|f_{i_1 \ldots i_n}(B_x(t))| = O(|C_x(t)|)$$

as t tends to 0 through S_x, where $f_{i_1 \ldots i_n} = (f_{i_1}, \ldots, f_{i_n})$, $1 \leq i_1 < \ldots < i_n \leq m$ and $|\cdot|$ is n-dimensional Lebesgue measure [15]. The main result is that $A(f) = I(f)$ for regular Sobolev mappings. Regular Sobolev mappings contain the ones considered above.

A more general approach, featuring the Jacobians themselves, rather than a Sobolev property imposed on the coordinate functions, was taken in [25]. Let d be the metric of (n-1)-continuity and let ρ be L_1-metric applied to Jacobians. Complete the space of smooth mappings with the $d+\rho$ metric. For f in the completion there is a generalized Jacobian $J_* f$. Sometimes J_* is the ordinary Jacobian J. This smaller class, using J, includes the Sobolev mappings properly. However, the mere existence of a summable Jacobian is not enough to put f in the above completion. An extension for the validity of $A(f) = I(f)$ is obtained in this setting. Details will not be given. We only mention the special fact that for continuous flat mappings the area measure, as defined by Cesari, and the measure given by the integral of J_* agree on the light part of the mapping.

It is not known just how much more general these approaches are than the approach given above where the coordinate functions are in Sobolev spaces.

8. Bounded variation was introduced by Jordan a hundred years ago as the condition for the length of a curve to be finite. By his decomposition of such a function as the difference of two monotone functions he immediately noted the functions in this class have everywhere convergent Fourier series. This dual application persists for higher dimensions. In the 1930's, Tonelli and Cesari applied functions of type BVC in two dimensions to the almost everywhere convergence of double Fourier series as well as to the surface area of a nonparametric surface. It was also noted that localization by rectangular sums holds for double series for functions of type BVC. The proof of almost everywhere convergence consists of two steps. With $V_x(f; y_0; [x_0, x_0 + \lambda])$ as notation for the variation of f in x on $[x_0, x_0 + \lambda]$ with $y = y_0$, the double Fourier series converges by rectangles at every point (x_0, y_0) such that, for every $\epsilon > 0$, there is a $\epsilon > 0$ such that $0 < \delta < \lambda$ implies

$$\frac{1}{2\delta} \int_{y_0-\delta}^{y_0+\delta} V_x(f; y_0; [x_0, x_0 + \lambda])dy < \epsilon ,$$

and three obviously similar conditions hold. Step 2 is that this condition holds almost everywhere, [3], [27].

For $n > 2$, there are functions in BVC, [28], and even in W_1^1, [12], for which the condition fails to hold anywhere using rectangles. It was found, however, [4], that for $n > 3$, if the variation functions are in L_p, $p > 1$, then the Fourier series of f converges almost everywhere by rectangular sums, and this result was extended in [31]. It seems that related methods should yield almost everywhere convergence using rectangular sums satisfying a parameter of regularity for every $f \in \text{BVC}$.

9. We turn now to localization. For $n = 1$, for any summable f, if $f = 0$ in an open set G then the Fourier series of f converges uniformly to zero on every compact $K \subset G$. It is well known that for two or more variables localization need not hold even for square sums. Indeed, for $n = 2$, there are everywhere differentiable functions for which localization by square sums does not hold. A paper which helped motivate this work, [21], treats the square (C,1) sums in n-dimensions. For each m, the average is taken of the $(m+1)^n$ partial sums $s_{i_0 i_1 \ldots i_m}$ of the multiple Fourier series of f, where $0 \leq i_1, \ldots, i_m \leq m$. Let $\{\sigma_m\}$ be this sequence of averages. Localization holds for square (C,1) sums if f equals zero on an open set G then $\{\sigma_m\}$ converges uniformly to zero on every compact subset of G.

The results are that localization holds by square (C,1) sums for every $f \in L_p$, $p \geq n-1$, but for every $p < n-1$ there is an $f \in L_p$ for which such localization does not hold. The proof uses only elementary functional analysis, i.e., the principle of uniform boundedness.

For rectangular (C,1) sums, it has been known for a long time that localization holds for continuous functions but not for all functions in L_p, for any $p \geq 1$.

10. For localization itself, not (C,1), the situation is different. Tonelli, [27], knew that for $n = 2$, localization by rectangular sums holds for every $f \in \text{BVC}$. Precise results have recently been obtained for $n = 2$ and will be discussed below. For $n > 2$, localization by both square and rectangular sums turns out to be in the Sobolev space, rather than the Cesari space, framework.

For any $f \in W_1^p$, $p \geq n-1$, localization by square sums holds. Moreover, for any $p < n-1$, there is an $f \in W_1^p$ for which localization by square sums does not hold. The proofs again only use the principle of uniform boundedness. It should be noted that the case $p = n-1$ is singular, since there are functions in W_1^{n-1} which are nowhere (n-1)-continuous. Indeed, [6], such functions may have a dense set of poles on almost every hyperplane parallel to a coordinate hyperplane. Although such functions have localization by square sums, there are functions $f \in W_1^{n-1}$ for which localization fails by rectangular sums.

However, if $f \in W_1^p$, $p > n$, then localization holds for rectangular sums. This proof uses the Morrey-Sobolev inequality, [23].

11. We define ΛBV for 2-dimensions. A function f is in ΛBV if there are equivalent g and h such that g is ΛBV as a function of y for almost all values of x and h is ΛBV as a function of x for almost all values of y, and if the two Λ-variation functions are summable, [19]. Now, if $f \in HBV$ it has localization by rectangular sums. If $\Lambda BV \not\subset HBV$, there is an $f \in \Lambda BV$ for which localization by square sums fails. It is remarkable that the conditions for square sums and rectangular sums are so close together.

BIBLIOGRAPHY

1. A. S. Besicovitch, "On the definition and value of the area of a surface", Quart. J. Math., Oxford 16 (1945), 86-102.

2. L. Cesari, "Sulle funzioni a variazione limitata", Annali Scuola. Norm. Sup. Pisa, Sec 2, 5 (1936), 299-313.

3. ________, "Sulle funzioni di due variabili a variazione limitata secondo Tonelli e sulla convergenza delle relative serie doppie di Fourier", Rend. del Sem. Mat. della Univ. di Roma, serie IV, I (1937), 277-294.

4. ________, "Sulle funzioni di più variabili generalmente a variazione limitata e sulla convergenza della relative serie multiple di Fourier", Comm. Pont. Acad. Sci 7 (1939), 171-197.

5. C. Goffman, "Lower semicontinuity and area functionals I., The nonparametric case", Rend. Circ. Mat. Palermo (1953), 203-235.

6. ________, "Nonparametric surfaces given by linearly continuous functions", Acta Math 103 (1960), 269-291.

7. ________, "Approximation of nonparametric surfaces of finite area", J. Math. Mech 12 (1963), 737-745.

8. ________, "A characterization of linearly continuous functions whose partial derivatives are measures", Acta Math 117 (1967), 165-190.

9. ________, "A remark in surface area", Y. W. Chen 60th birthday volume, 1970, Taipei.

10. ________, "An example in surface area", J. of Math. and Mech. 19 (1969), 321-326.

11. ________, "Everywhere convergence of Fourier series", Indiana U. Math. J., 20 (1970), 102-112.

12. ________, "An example of Torrigiani related to multiple Fourier series", Mich. Math J., 19 (1972), 285-287.

13. ________, "Coordinate invariance of linear continuity", Arch. for Rat. Mech. and Anal, 20 (1965), 153-162.

14. ________ and F. C. Liu, "Discontinuous mappings and surface area", Proc. London Math. Soc., 20 (1970), 237-248.

15. ________, "The area formula for Sobolev mappings", Indiana Math. J., 25 (1976), 871-876.

16. ________, "Localization of square sums of multiple Fourier series", Studia Math. 44 (1972), 61-69.

17. ________, "Lusin type theorems for functions of bounded variation", Real Anal. Exch. 5 (1979), 261-262.

18. ________, "Derivative measures", Proc. Amer. Math. Soc. 78 (1980), 218-220.

19. C. Goffman and D. Waterman, "The localization principle for double Fourier series", Studia Math., 69 (1980), 41-57.

20. C. Goffman and W. Ziemer, "Higher dimensional mappings for which the area formula holds", Ann. Math. 92 (1970), 482-489.

21. S. Igari, "On the localization property of multiple Fourier series", J. Approx. Theory, 1 (1968), 182-188.

22. F. C. Liu, "Approximation extension type property of continuous functions of bounded variation", J. Math. Mech 19 (1970), 207-218.

23. ________, "On the localization of rectangular partial sums for multiple Fourier series", Proc. Amer. Math. Soc. 34 (1972), 90-96.

24. J. H. Michael, "The equivalence of two areas for non-parametric discontinuous surfaces", Illinois J. of Math. 7 (1963), 59-78.

25. T. Nishiura and J. C. Breckenridge, "Differentiation, integration, and Lebesgue area", Indiana U. Math. J. 26 (1977), 515-536.

26. J. Serrin, "On the differentiability of functions of several variables", Arch. Rat. Mech. Anal. 7 (1961), 359-372.

27. L. Tonelli, Sulle serie doppie di Fourier", Ann. Scuoli Norm. Sup. Pisa 6 (1937), 315-326.

28. G. Torrigiani, "Sulle funzioni di piu variabili a variazione limitata", Riv. Mat. Univ. Parma 1 (1950), 59-83.

29. D. Waterman, "On convergence of Fourier series of functions of generalized bounded variation", Studia Math. 44 (1972), 107-117.

30. C. Goffman and D. Waterman, "A characterization of functions whose Fourier series converge everywhere for all changes of variable", J. London Math. Soc. 10 (1975), 69-74.

31. Jau D. Chen, "A theorem of Cesari on multiple Fourier series", Studia Math. 49 (1973), 69-80.

DEPARTMENT OF MATHEMATICS
PURDUE UNIVERSITY
WEST LAFAYETTE, INDIANA 47907

Contemporary Mathematics
Volume 42, 1985

AN UNUSUAL DESCRIPTIVE DEFINITION OF INTEGRAL

B. Bongiorno, D. Preiss

ABSTRACT. We study those continuous functions fulfilling the Luzin condition (N) which are uniquely (up to an additive constant) determined by their derivative. As a by-product we get an unusual definition of an integral which turns out to be equivalent to the Denjoy-Chincin integral.

This work arose from two independent, but very much connected observations. Firstly, in Saks [3, p.285] one already finds that, whenever f and g are real-valued functions defined on an interval such that g is absolutely continuous, f is continuous and has the Luzin property (N) (i.e. maps Lebesgue zero sets onto Lebesgue zero sets) and $f' = g'$ a.e., then $f - g$ is a constant. Absolute continuity of g can be replaced by ACG_* (this is noted in [3] and proved in [2]) and even by ACG, in which case the statement remains valid with ordinary derivative replaced by approximate derivative [2]. On the other hand, absolute continuity and Luzin property cannot be replaced by continuity and Luzin property [2]. Hence, it seems to be natural to ask to which classes of functions these results extend.

Secondly, we consider the way in which descriptive definitions of integrals are obtained. Usually, when defining the indefinite integral of ϕ, we give some condition which chooses among all primitives of ϕ (i.e. among all functions whose (approximate) derivative equals ϕ a.e.) at most one (up to an additive constant). In view of the results mentioned above another possibility occurs. We may try to invent some property of primitives and then to define the indefinite integral of ϕ as a primitive possessing this property, provided that it exists and is determined uniquely up to an additive constant. One natural candidate for a condition to be imposed on primitives is continuity and Luzin property. With this definition, the family of integrable functions contains Denjoy integrable functions (or Denjoy-Chincin integrable, if approximate derivatives are used). Since it would be rather interesting to get an extension of the Denjoy-Chincin integral in such a

1980 Mathematics Subject Classification. 26A39.

simple way, we ask, whether this really gives an extension. This question is easily seen to be equivalent to that already asked above.

Now, we may summarize our answers. We shall consider real-valued functions defined on an interval $[a,b]$ and we shall denote by (N) the family of all functions which are continuous and have the Luzin property. For such notions like VBG etc. the reader is referred to [2].

Theorem 1. Assume that $f \in (N)$ and that f is differentiable (approximately differentiable, respectively) a.e. Then f is ACG if and only if for every function $g \in (N)$ with $g' = f'$ a.e. ($g'_{ap} = f'_{ap}$ a.e., respectively) the difference $g - f$ is constant.

Theorem 2. If ϕ is a function defined on $[a,b]$, then ϕ is Denjoy-Chincin integrable iff there is exactly one (up to an additive constant) function $f \in (N)$ such that $f'_{ap} = \phi$ a.e.

Theorem 3. If ϕ is a function defined on $[a,b]$, then there is exactly one (up to an additive constant) function $f \in (N)$ such that $f' = \phi$ a.e. if and only if ϕ is Denjoy-Chincin integrable and its indefinite integral has the ordinary derivative a.e.

These results follow immediately from the following two propositions. The first of them has practically been proved in [2], the second is the main result of this work.

Proposition 1. If $f \in (N)$, g is ACG on $[a,b]$ and $f'_{ap} \geqq g'_{ap}$ a.e., then $f - g$ is nondecreasing.

Proof. If not, we find $a \leqq x < y \leqq b$ such that $f(x) - g(x) > f(y) - g(y)$. Let $h(t) = \max\{(f(s) - g(s));\ s \in [t,y]\}$ for $t \in [x,y]$. Then $g + h$ is VBG on $[x,y]$ since g is VBG and h is monotone and continuous. Since h is constant on every interval contiguous to $\{t \in [x,y];\ g(t) + h(t) = f(t)\}$, the function $g + h$ has the Lusin property and $(g + h)'_{ap} \geqq g'_{ap}$ a.e. Since VBG and (N) imply ACG, the function $g + h$ and, consequently, also the function h are ACG. Since $h'_{ap} \geqq 0$ a.e., h is nondecreasing, which contradicts its definition.

Proposition 2. If $f \in (N)$ and if f is not ACG, then there is a continuous non-constant function g such that $g' = 0$ a.e. and $f + g \in (N)$.

Using the fact that f is ACG iff it is VBG and has the Luzin property, we see that this proposition is an immediate consequence of the following statement.

Proposition 3. If f is a continuous function defined on $[a,b]$ and if f is not VBG, then there are a closed set $M \subset [a,b]$ of measure zero and a monotone, non-constant, continuous function g on $[a,b]$ which is constant on each component of $[a,b]\setminus M$ and for which $|(f+g)(M)| = 0$.

Proof. Let G be the union of all open intervals I such that f is VBG on $I \cap [a,b]$. Since G can be written as a countable union of such intervals, f is VBG on $G \cap [a,b]$. It follows that the set $Q = [a,b] - G$ is non-empty and perfect and that f is not VBG on any portion of Q. Passing to a portion of Q and changing f to $-f$ if necessary, we may also assume that the negative part of the variation of f is infinite on each portion of Q. This means that we need to prove the statement of the proposition under the following assumptions.

f is a continuous function on $[a,b]$, $Q \subset [a,b]$ is nonempty, perfect and such that for every open interval T intersecting Q and for every $C > 0$ there are points $u_1 < u_2 < \ldots < u_{2p-1} < u_{2p}$ in $T \cap Q$ such that $f(u_{2i-1}) > f(u_{2i})$ and

$$\sum_{i=1}^{p} (f(u_{2i-1}) - f(u_{2i})) > c.$$

We shall keep these assumptions in the following lemmas.

Lemma 1. If $u \in Q \cap (a,b)$, $\varepsilon > 0$, $A,B \in \mathbb{R}$ and $A < B$ then there are a finite family $I_1, I_2, \ldots, I_{2p}$ of closed subintervals of (a,b) and a nondecreasing continuous functions h on $[a,b]$ such that

(i) $\operatorname{diam}(\{u\} \cup \bigcup_{i=1}^{2p} I_i) < \varepsilon$,

(ii) $Q \cap \operatorname{Int}(I_i) \neq \emptyset$ $(i = 1, \ldots, 2p)$,

(iii) h is constant on each component of $[a,b] - \bigcup_{i=1}^{2p} I_i$ and is linear and increasing on each I_i,

(iv) $h(a) = A$, $h(b) = B$,

(v) $\sum_{i=1}^{2p} |f(I_i) + h(I_i)| \leqq B - A + \varepsilon$, and

(vi) $|\bigcup_{i=1}^{2p} [f(I_i) + h(I_i)]| \leqq \frac{1}{2}(B - A) + \varepsilon$,

where we denoted $X + Y = \{x + y \,;\, x \in X,\, y \in Y\}$ for $X, Y \subset \mathbb{R}$.

Proof. Let $T \subset (a,b)$ be an open interval such that $u \in T$, $|T| < \varepsilon$

and $|f(T)| < \frac{1}{10}\min(\varepsilon, B-A)$. Let $u_1 < u_2 < \dots < u_{2p}$ be elements of $T \cap Q$ such that $f(u_{2i-1}) > f(u_{2i})$ and

$$M = \sum_{i=1}^{p-1} (f(u_{2i-1}) - f(u_{2i})) < \tfrac{1}{2}(B - A) \leqq \sum_{i=1}^{p} (f(u_{2i-1}) - f(u_{2i})).$$

We note that $p > 1$ and $\tfrac{1}{2}(B - A) - M < \frac{1}{10}\varepsilon$.

Let $I_1, I_2, \dots, I_{2p}$ be disjoint, closed intervals contained in T such that $u_i \in \mathrm{Int}(I_i)$ and $|f(I_i)| < \frac{\varepsilon}{10p}$. Define a continuous function h on $[a,b]$ such that (iii) holds, $h(a) = A$, $|h(I_{2i-1})| = |h(I_{2i})| = f(u_{2i-1}) - f(u_{2i})$ for $i = 1, 2, \dots, p-1$, and $|h(I_{2p-1})| = |h(I_{2p})| = \tfrac{1}{2}(B - A) - M$. Then $h(b) = B$ and therefore (i) - (iv) hold.

Clearly $|f(I_i) + h(I_i)| = |f(I_i)| + |h(I_i)| \leqq |h(I_i)| + \frac{\varepsilon}{10p}$, which shows (v). Moreover, for $i = 1, \dots, p-1$
$f(u_{2i}) + \min(h(I_{2i})) = f(u_{2i}) + \max(h(I_{2i-1})) = f(u_{2i}) + \min(h(I_{2i-1})) + h(I_{2i-1}) = f(u_{2i-1}) + \min(h(I_{2i-1}))$, and also since $h(I_{2i-1}) = h(I_{2i})$, $f(u_{2i}) + \max(h(I_{2i})) = f(u_{2i-1}) + \max(h(I_{2i-1}))$. Hence $\{f(u_{2i-1})\} + h(I_{2i-1}) = \{f(u_{2i})\} + h(I_{2i})$ for $i = 1, \dots, p-1$. Thus $[f(I_{2i-1}) + h(I_{2i-1})] \cup [f(I_{2i}) + h(I_{2i})]$ is a subset of the $\frac{\varepsilon}{10p}$ - neighborhood of $f(u_{2i}) + h(I_{2i})$ $(i = 1, \dots, p-1)$ and consequently

$$\left|\bigcup_{i=1}^{2p} [f(I_i) + h(I_i)]\right| \leqq \sum_{i=1}^{p-1} \left(|f(u_{2i}) + h(I_{2i})| + \frac{\varepsilon}{5p}\right) + |h(I_{2p-1})| + \frac{\varepsilon}{5p}$$
$$+ |h(I_{2p})| + \frac{\varepsilon}{5p} \leqq \sum_{i=1}^{p-1} h(I_{2i}) + \varepsilon < \tfrac{1}{2}(B - A) + \varepsilon.$$

<u>Lemma 2</u>. If $y_0 < y_1 < \dots < y_r$, $\varepsilon > 0$, $I \subset [a,b]$ is a closed interval such that $Q \cap \mathrm{Int}(I) \neq \emptyset$ and g is an increasing linear function on I such that $f(I) + g(I) \subset [y_0, y_r]$, then there is a finite family $T_1, \dots, T_q$ of disjoint, closed subintervals of I and a continuous, nondecreasing function h on I such that

(i) $\left|\bigcup_{i=1}^{q} T_i\right| < \varepsilon$,

(ii) $Q \cap \mathrm{Int}(T_i) \neq \emptyset$,

(iii) h is constant on each component of $I - \bigcup_{i=1}^{q} T_i$ and is linear and increasing on each T_i,

(iv) $h = g$ at the end-points of I,

(v) for each $j = 1, \ldots, r$

$$|(y_{j-1},y_j) \cap \bigcup_{i=1}^{q}[f(T_i) + h(T_i)]| \leq \tfrac{1}{2}(y_j - y_{j-1}) + \varepsilon$$

(vi) for each $j = 1, \ldots, r$

$$\sum_{i=1}^{q} |(y_{j-1},y_j) \cap [f(T_i) + h(T_i)]| \leq (y_j - y_{j-1}) + \varepsilon, \text{ and}$$

(vii) for each $i = 1, \ldots, q$

$$|f(T_i) + h(T_i)| \leq \varepsilon.$$

Proof. Put $\delta = \frac{\varepsilon}{8}$. Let $v \in Q \cap \mathrm{Int}(I)$ and let $z_0 < z_1 < \ldots < z_s$ be such that $f(v) + g(I) = [z_0,z_s]$ and $\max(z_k - z_{k-1}) < \delta$. We find a closed interval $\hat{I}$ contained in I such that $Q \cap \mathrm{Int}(\hat{I}) \neq \emptyset$ and $|f(\hat{I})| < \delta$, and we choose a sequence $u_1 < u_2 < \ldots < u_s$ of points of $Q \cap \mathrm{Int}(\hat{I})$.

Let $\Delta > 0$ be such that $\mathrm{dist}([u_1,u_s], \mathbb{R} - \hat{I}) > \Delta$, $(u_{i+1} - u_i) > 2\Delta$, and $2s\Delta < \varepsilon$.

For each $j = 1, \ldots, s$ we use Lemma 1 with u, ε, A and B replaced by u_j, Δ, $(z_{j-1} - f(v))$ and $(z_j - f(v))$, respectively, to construct families $I_1^j, I_2^j, \ldots, I_{2p_j}^j$ of disjoint, closed intervals and functions h^j with properties (1.i) - (1.vi).

From (1.i) and (1.iv) we see that the family $T_1, T_2, \ldots T_q$ (where $q = 2p_1 + \ldots + 2p_s$) of all these intervals is disjoint and that there is a continuous function h fulfilling (iii), (iv) and such that $h = h^j$ on on each I_i^j.

Now, (i) follows immediately from (1.i) and $2s\Delta < \varepsilon$, and (ii) follows from (1.ii).

To prove (v) and (vi), we first note that

$$(*) \quad f(I_i^j) + h(I_i^j) = f(I_i^j) + h^j(I_i^j)$$
$$\subset f(I_i^j) + [z_{j-1} - f(v),\ z_j - f(v)] \subset [z_{j-1} - \delta,\ z_j + \delta].$$

Hence

$$|(y_{j-1},y_j) \cap \bigcup_{i=1}^{q} [f(T_i) + h(T_i)]| \leq \Sigma^{(j)} |\bigcup_{i=1}^{2p_k} [f(I_i^k) + h^k(I_i^k)]|,$$

(where $\Sigma^{(j)}$ extends over those $k = 1, \ldots, s$ for which $[z_{k-1} - \delta, z_k + \delta] \cap (y_{j-1},y_j) \neq \emptyset$) which can be estimated using (1.vi), as

$$\leq [\Sigma^{(j)} \tfrac{1}{2}(z_k - z_{k-1})] + \Delta s$$
$$\leq \tfrac{1}{2}(y_j - y_{j-1}) + \delta + \max_k(z_k - z_{k-1}) + \tfrac{1}{2}\varepsilon$$
$$\leq \tfrac{1}{2}(y_j - y_{j-1}) + \varepsilon.$$

Similarly we prove (vi).

$$\sum_{i=1}^{q} |(y_{j-1},y_j) \cap [f(T_i) + h(T_i)]|$$
$$\leq \Sigma^{(j)} \sum_{i=1}^{2p_k} |f(I_i^k) + h^k(I_i^k)|$$
$$\leq [\Sigma^{(j)} (z_k - z_{k-1})] + \Delta s$$
$$\leq (y_j - y_{j-1}) + 2\delta + 2\max_k(z_k - z_{k-1}) + \tfrac{1}{2}\varepsilon$$
$$\leq (y_j - y_{j-1}) + \varepsilon.$$

Finally we note that (vii) follows immediately from (*).

Lemma 3. There are a closed set $M \subset [a,b]$ with $|M| = 0$ and a non-decreasing, non-constant, continuous function g which is constant on each component of $[a,b] - M$ and for which $|(f + g)(M)| = 0$.

Proof. Let K be the range of the function $f(t) + \frac{t-a}{b-a}$ on $[a,b]$. We shall consider K as a probability space, the probability being given by the normalized Lebesque measure. Whenever J is a sub-σ-algebra of Lebesque measurable sets, we denote by $P(\cdot|J)$ and $E(\cdot|J)$ the conditional probability and the conditional expectation, respectively. By X_A we shall denote the characteristic function of the set A.

By induction we construct a nondecreasing sequence p_n of natural numbers, a sequence g_n of nondecreasing, continuous function on $[a,b]$, sequences $I_1^n, \ldots, I_{p_n}^n$ of disjoint closed subintervals of $[a,b]$ and a sequence J_n of σ-algebras on K with the following properties.

(i) $p_o = 1$, $I_1^o = [a,b]$, $g_o(t) = \frac{t-a}{b-a}$ and J_o is the smallest σ-algebra on K.

(ii) $I_j^n = I_{j+1}^{n-1}$ for $1 \leq j \leq p_{n-1} - 1$ and $n = 1, 2, \ldots$

(iii) $\bigcup_{j=p_{n-1}}^{p_n} I_j^n \subset I_1^{n-1}$ for $n = 1, 2, \ldots$

(iv) $|\bigcup_{j=p_{n-1}}^{p_n} I_j^n| \leq 2^{-n} p_{n-1}^{-1}$ for $n = 1, 2, \ldots$

(v) $Q \cap \operatorname{Int} I_j^n \neq \emptyset$ for $n = 0, 1, \ldots$ and $j = 1, \ldots, p_n$.

(vi) g_n is constant on each component of $[a,b] - \bigcup_{j=1}^{p_n} I_j^n$ and is linear and increasing on each I_j^n $(n = 0, 1, \ldots)$.

(vii) $g_n = g_{n-1}$ on $[a,b] - I_1^{n-1}$ $(n = 1, 2, \ldots)$.

(viii) $|f(I_j^n) + g_n(I_j^n)| < 2^{-n}$ for $n = 1, 2, \ldots$ and $j = p_{n-1}, p_{n-1} + 1, \ldots p_n$.

(ix) Each J_n is a finite σ-algebra of subsets of K generated by some finite partition of K into closed non-overlapping intervals with length $\leq 2^{-n}|K|$.

(x) $J_n \supset J_{n-1}$.

(xi) The sets $f(I_j^n) + g_n(I_j^n)$ are J_n - measurable.

(xii) $P(\bigcup_{j=p_{n-1}}^{p_n} [f(I_j^n) + g_n(I_j^n)] \mid J_{n-1}) \leq \frac{3}{4}$ for $n = 1, 2, \ldots$

(xiii) $E(\sum_{j=1}^{p_n} \chi_{f(I_j^n) + g_n(I_j^n)} \mid J_{n-1}) \leq \bar{2}^n + \sum_{j=1}^{p_{n-1}} \chi_{f(I_j^{n-1}) + g_{n-1}(I_j^{n-1})}$ for $n = 1, 2, \ldots$

For $n = 0$ we use (i) as a definition, since it is easy to see that the remaining conditions hold.

If all our objects have already been defined for $n-1$, let $y_o < y_1 < \ldots < y_r$ be a partition of $f(I_1^{n-1}) + g_{n-1}(I_1^{n-1})$ which generates the restriction of J_{n-1} to $f(I_1^{n-1}) + g_{n-1}(I_1^{n-1})$. We use Lemma 2 with $I = I_1^{n-1}$, $\varepsilon = 2^{-n-1} \min(p_{n-1}^{-1}, (y_1 - y_o), \ldots, (y_r - y_{r-1}))$, and $g = g_{n-1}$ to construct the family $T_1, \ldots, T_q$ and the function h with the properties described there.

We define $p_n = p_{n-1} - 1 + q$, $I_i^n = I_{i+1}^{n-1}$ for $i = 1, \ldots, p_{n-1} - 1$,

$I_i^n = T_{i-p_{n-1}+1}$ for $i = p_{n-1}, p_{n-1}+1, \ldots, p_n$,

$g_n = g_{n-1}$ on $[a,b] - I_1^{n-1}$, and

$g_n = h$ on I_1^{n-1}.

Then (i), (ii), (iii) and (vii) follow immediately from the definitions, (iv) is implied by (2.i), (v) by (2.ii), (vi) by (2.iii) and (viii) by (2.vii).

Finally, we define J_n such that (ix), (x) and (xi) hold.

To prove (xii), we note that, for $j \geq p_{n-1}$, $f(I_j^n) + g_n(I_j^n) \subset f(I_1^{n-1}) + g_{n-1}(I_1^{n-1}) = [y_o, y_r]$, hence it suffices to show that for each

$$j = 1, 2, \ldots, r \quad |(y_{j-1}, y_j) \cap \bigcup_{i=p_{n-1}}^{p_n} [f(I_i^n) + g_n(I_i^n)]| \leq \tfrac{3}{4}(y_j - y_{j-1}),$$

which follows from (2.v).

To prove (xiii), we note that, since $\sum_{j=1}^{p_{n-1}-1} \chi_{f(I_j^n) + g(I_j^n)}$ is J_{n-1}

measurable, it suffices to show that

$$E(\sum_{i=p_{n-1}}^{p_n} \chi_{f(I_i^n) + g_n(I_i^n)} | J_{n-1}) \leq 2^{-n} + \chi_{f(I_1^{n-1}) + g_{n-1}(I_1^{n-1})}.$$

Similiarly as in showing (xii) we reduce this to proving that for each

$j = 1, 2, \ldots, r$

$$\int_{(y_{j-1}, y_j)} \sum_{i=p_{n-1}}^{p_n} \chi_{f(I_i^n) + g_n(I_i^n)} \leq (1 + 2^{-n})(y_j - y_{j-1}),$$

which follows from (2.vi).

Having thus constructed these approximating sequences, we first note that (ii), (iii) and (iv) imply that $|\bigcup_{j=1}^{p_{n+p_n}} I_j^{n+p_n}| \leq 2^{-n}$ and that

$\bigcup_{j=1}^{p_n} I_j^n \supset \bigcup_{j=1}^{p_{n+1}} I_j^{n+1}$. Hence $M = \bigcap_{n=0}^{\infty} (\bigcup_{j=1}^{p_n} I_j^n)$ is a closed subset of $[a,b]$ with $|M| = 0$.

Similarly, from (ii), (iii) and (viii) we deduce that

$$\lim_{n\to\infty} \max_{j=1,\ldots p_n} |f(I_j^n) + g_n(I_j^n)| = 0.$$

Since g_m are monotone, (ii), (iii), (vi) and (vii) imply

$$\max_{t\in[a,b]} |g_m(t) - g_n(t)| \leq \max_{j=1,\ldots p_n} |g_n(I_j^n)|$$

for $m \geq n$. Thus the above observation shows that the sequence g_n is uniformly convergent. Let g denote its limit.

To finish the proof, it remains to show that $|(f + g)(M)| = 0$ (other statements are easy). Let

$$\psi_n = \sum_{j=1}^{p_n} X_{f(I_j^n) + g_n(I_j^n)}$$

Since $(f + g)(M) \subset \bigcup_{j=1}^{p_n} [f(I_j^n) + g(I_j^n)] \subset \{x; \psi_n(x) \geq 1\}$ for each n, the proof will be finished by showing that ψ_n converges to zero a.e. To prove this, we first note that (xiii) shows that $\psi_n + 2^{-n}$ is a nonnegative submartingale, hence $\psi_n + 2^{-n}$ (and, therefore, ψ_n) converges a.e. [1]. Assume that the limit ψ of ψ_n is positive on a set A of positive measure. We may also assume that ψ_n converge uniformly to ψ on A. Since ψ_n attain only integer values, there is a positive integer, say K, and n_0 such that $\psi_n(x) = K$ for each $n \geq n_0$ and each x belonging to some set B of positive measure. Let $z \in B$ be a point of density of B, and let $\delta > 0$ be such that $|L - B| \leq \frac{1}{8}|L|$ for every interval L containing z such that $|L| \leq \delta$. We find an $n \geq n_0$ such that the partition generating J_n contains an interval L with $z \in L$ and $|L| \leq \delta$. Since ψ_n is J_n-measurable, there is $j \in \{1, \ldots, p_n\}$ such that $f(I_j^n) + g_n(I_j^n) \supset L$. Let q be the smallest among such j. Let $m = n + q - 1$. Then $I_1^m = I_q^n$ and $g_m = g_n$ on I_1^m, hence $f(I_1^m) + g_m(I_1^m) \supset L$. Consequently, for every $x \in B \cap L$

$$K-1 = \sum_{j=2}^{p_m} X_{f(I_j^m) + g_m(I_j^m)}(x) = \sum_{j=1}^{p_m-1} X_{f(I_j^{m+1}) + g_{m+1}(I_j^{m+1})}(x)$$

and in view of (xii),

$$|L \cap \bigcup_{j=p_m}^{p_{m+1}} [f(I_j^{m+1}) + g_{m+1}(I_j^{m+1})]| \leq \frac{3}{4} L.$$

Hence the set $(B \cap L) - \bigcup_{j=p_m}^{p_{m+1}} [f(I_j^{m+1}) + g_{m+1}(I_j^{m+1})]$ is nonempty (its measure is $\geq \frac{1}{8} L$) and $\psi_{m+1} = K - 1$ on this set, which is impossible since

$\psi_m = K$ on B. This contradiction finishes the proof of Lemma 3 as well as of Proposition 3.

Acknowlegement: The main part of this work was done in the framework of activities of D. Preiss as visiting professor at University of Palermo sponsored by C.N.R. of Italy, and was finished during his participation in the Special Year in Real Analysis at the University of California, Santa Barbara.

BIBLIOGRAPHY

1. L. Breiman, Probability, Addison-Wesley, 1968.
2. J. Foran, "Differentiation and Luzin's condition (N)", Real Analysis Exchange 3(1977-78), no. 1, 34-37.
3. S. Saks, Theory of the Integral, Dover Publications, New York, 1937.

DEPARTMENT OF MATHEMATICS
UNIVERSITY OF PALERMO
via ARCHIRAFI 34
PALERMO, ITALY

and

DEPARTMENT OF MATHEMATICS
CHARLES UNIVERSITY
SOKOLOVSKA 83
PRAGUE 8, CZECHOSLOVAKIA

Current Address:
Department of Mathematics
University of California
Santa Barbara, CA 93106

Contemporary Mathematics
Volume 42, 1985

PATH DERIVATIVES: A UNIFIED VIEW OF CERTAIN GENERALIZED DERIVATIVES

A. M. Bruckner,[1] R. J. O'Malley,[1] and B. S. Thomson[2]

ABSTRACT. The limit of a difference quotient, $(F(y)-F(x))/(y-x)$ may be considered as $y \to x$ through a collection of sets E_x known as a path system. Four path systems are described and it is shown why certain properties of derivatives hold also for generalized derivatives.

It is natural to desire one framework within which such generalized derivatives as the approximate, Peano, or Dini can be expressed and which reveals why various properties of derivatives are inherited by the generalizations. This paper constructs such a framework for at least those cases where the generalized derivative of a function F at a point x can be viewed as:

$$\lim_{\substack{y \to x \\ y \in E_x}} \frac{F(y)-F(x)}{y-x}$$

for appropriate choices of sets, E_x, where x belongs to E_x and E_x has x as a limit point. When we have a choice of such sets E_x, for each $x \in R$, the collection is called a path system and denoted E. The corresponding differentiation theory uses notation and definitions consistent with the classical theory. For example, $F_E'(x)$ is the path derivative of F at x relative to E and $\overline{F}_E'(x)$ is the corresponding upper extreme path derivate. Since it is our purpose here to present just enough material to indicate the thrust and core of the theory we will have to delete many of the theorems contained in [1] and only outline some of the proofs. In addition we will restrict our attention to situations where our functions have finite path derivatives everywhere. (The interested reader should also review [7] and [8].)

To fix the concepts we label four natural path systems. That is, $E=\{E_x : x \in R\}$ is a path system of

a) ordinary type if E_x is a neighborhood of x,

1980 Mathematics Subject Classification. 26A24.

[1]The work of these authors was supported in part by grants from the National Science Foundation.

[2]The work of this author was supported in part by a grant from the National Sciences and Engineering Research Council of Canada.

b) (1,1) density type if each E_x has density 1 at x from the left and right,

c) congruent type if $E_x = E_0 + x$,

d) qualitative type if each E_x is residual in some neighborhood of x.

It should be easy for the reader to formulate various modifications of the above four to one sided types or other density types.

It is established in the paper that much of the structure possessed by the path derivative and its primitive is dependent upon the geometry of the system E. In particular two different types of characteristic seem to play an essential role. First, we have "thickness" criteria such as bi) bilateral type if each E_x has x as a bilateral limit point, np) nonporous type if each E_x has porosity 0 at x.

The porosity of a set E at a point x from the right is defined, [2], [3], as the value:

$$\limsup_{r \to 0^+} \frac{\ell(x,r,E)}{r}$$

where $\ell(x,r,E)$ denotes the length of the largest open interval in the set $(x,x+r) \cap (R \setminus E)$. Porosity 0 at x means both right and left porosity 0. Basically, porosity 0 forces E to have "relatively small" gaps between its points as we get closer to x. It should be noted that of the four examples of path system listed earlier a, b and d are non-porous type.

Second, in addition to the "thickness" it was found that various types of intersection conditions between E_x and E_y are involved. More precisely a path system is said to satisfy the type of intersection condition listed below, if there is associated with E a positive function δ on R, so that whenever $0<y-x<\min\{\delta(x),\delta(y)\}$, the paths E_x and E_y intersect in the stated fashion:

i) intersection condition: $E_x \cap E_y \cap [x,y] \neq \phi$,

ii) internal intersection condition: $E_x \cap E_y \cap (x,y) \neq \phi$,

iii) external intersection condition: $E_x \cap E_y \cap (y,2y-x) \neq \phi$ and $E_x \cap E_y \cap (2x-y,x) \neq \phi$,

iv) external intersection condition, parameter m $E_x \cap E_y \cap (y,(m+1)y-mx) \neq \phi$ and $E_x \cap E_y \cap ((m+1)x-my,x) \neq \phi$,

v) one-sided external intersection (parameter m): same as iii) and iv) but only one of the intersections need be nonempty.

It should be noted that (1,1) density type path systems satisfy all these conditions.

Now we will try and indicate how these various concepts combine. Throughout we will assume we have a fixed function $F:R \to R$ which with respect to a path system E has a finite path derivative F'_E. It will help the presentation if we think first about a (1,1) density system, that is, approximately differentiable functions and approximate derivatives.

It was known that:

1) F is Baire 1 but not necessarily continuous. In fact, F is [ACG] in the sense that there is a sequence of closed sets X_n such that F is absolutely continuous relative to X_n and $\bigcup_{n=1}^{\infty} X_n = R$.

2) $F_E' = F_{ap}'$ is in Zahorski's $\mathcal{M}_3$ class and in particular F_{ap}' is Baire 1, Darboux and possesses the Denjoy-Clarkson property.

3) If $F_{ap}' \geq 0$, then F is nondecreasing. Further if F is nondecreasing then F is actually differentiable. In turn, these two facts yield that if F_{ap}' is bounded then F is differentiable and absolutely continuous.

4) There is a dense open set U such that, on each component of U, F is differentiable. It is also known that if M is given and F_{ap}' takes on values M and $-M$, then there is some component of U on which F' takes on both M and $-M$ as values.

We consider to what extent these facts can be generalized to an arbitrary path system E.

We have:

1*) If E satisfies any of the intersection conditions i) through v) then F will be Baire 1 and in fact [ACG]. The proof of this fact makes use initially of what is called a δ-decomposition where δ is the function associated with E for the appropriate intersection condition. It involves the sets

$$Y_{mj}=\{x:\delta(x) > \tfrac{1}{m}\} \cap [\tfrac{j}{m}, \tfrac{(j+1)}{m}],\ m=1,2,\ldots \text{ and } j=0,\pm1,\pm2,\ldots .$$

When two points x and y belong to the same Y_m, the intersection condition involved can be applied to E_x and E_y.

Next we use the fact that F_E' is finite to consider sets X_n such that $\bigcup_{n=1}^{\infty} X_n = R$ where

$$X_n = \{x: -n < \frac{f(x)-f(y)}{x-y} < n \text{ when } y \in E_x \text{ and } |x-y| < \tfrac{1}{n}\} .$$

For a fixed n, consider $X_{nmj}=X_n \cap Y_{mj}$ for $m=n,n+1,\ldots$ and $j=0,\pm1,\pm2,\ldots$. It can be shown that F will be Lipschitz on each X_{nmj} and further that this behavior will extend to the closure of X_{nmj}. Thus F is [ACG].

It should be noted that being [ACG] forces F to be approximately differentiable almost everywhere. This doesn't of itself automatically imply that F_E' is the approximate derivative of F almost everywhere. However, it is true that $F_E' = F_{ap}'$ a.e. if E has the intersection condition i).

We consider next when F_E' will be in the various Zahorski classes 2 and 3 and also the subject of item 3.

For F_E' to be Baire 1 it is not sufficient for E to have any intersection property. However, either the external intersection condition or external intersection condition parameter m is sufficient.

If we have F_E' in Baire class 1 by some method, then F_E' will become Darboux if E is bilateral and it is easy to see that this bilateral condition is necessary by considering such examples as $F(x)=|x|$.

For the M_2 class or the Denjoy-Clarkson property we find that a combination of conditions help. For example if E has both the intersection condition and external intersection condition and is bilateral, then F_E' will have the Denjoy-Clarkson property. This result rests basically on the principle that under such a set of conditions if $F_E' \geq 0$ almost everywhere in an interval, then F is nondecreasing on the interval.

However, unlike the case with approximate derivatives, if F is nondecreasing we do not automatically have that F_E' is actually the derivative of F. It is this that prevents us from being guaranteed that F_E' is in Zahorski's $\mathcal{M}_3$ class under these conditions.

It is here that we first see a use for the concept of nonporosity. In general we have the following pointwise result. If F is monotonic and E_X has porosity zero at x, then $\underline{F}_E'(x) = \underline{F}'(x)$ and $\overline{F}_E'(x) = \overline{F}'(x)$. When this is applied to a monotone E-differentiable function with a nonporous system we obtain that the E-derivative is actually a derivative. Then we have results such as:

If E is nonporous and satisfies the intersection condition and external intersection condition, then F_E' is in Zahorski's $\mathcal{M}_3$ class and in fact every set of the form $\{x: \alpha < F_E'(x) < \beta\}$ is of class $\mathcal{M}_3$.

Further: Under the same hypothesis there is a dense open set U such that F is differentiable at every point of U which improves the situation mentioned in 1*.

Finally in connection with 4 we have

4*) Suppose E is nonporous and has the intersection property and F_E' is Baire 1, then if F_E' attains the values M and $-M$ on an interval I_0, there is a subinterval I of I_0 on which F is differentiable and F' attains the values M and $-M$.

A number of applications of this result are discussed in [1].

We finish our summary by pointing out two open questions associated with path derivatives. The concept of selective differentiation was introduced in [4] and further developed in [5], [6]. It is quite easy to show that if F has a path derivative F_E' via a system which is bilateral and has the internal intersection property then F_E' is a selective derivative of F. However the converse still is an open question. Finally, it is known that if a function F has a $(k+1)^{th}$ Peano derivative, then F_k' has F_{k+1}' as a path derivative via a system which is nonporous. Yet it is not known whether there is a path system having the intersection property which achieves the same result.

REFERENCES

[1] A. M. Bruckner, R. J. O'Malley, B. S. Thomson, Path derivatives: a unified view of certain generalized derivatives, to appear in Trans. A.M.S.

[2] E. P. Dolzenko, Boundary properties of real functions, Math. USSR-Izv.: (1967), 1-12.

[3] M. J. Evans, P. D. Hunke, The equality of unilateral derivates Proc. A.M.S. 79 (1980), 609-613.

[4] R. J. O'Malley, Selective derivates, Acta Math. Acad. Sci. Hung. 29 (1977), 77-97.

[5] _______, Selective derivatives and the M_2 or Denjoy, Clarkson properties, Acta Math. Acad. Sci. Hung., 36 (1980), 195-199.

[6] _______, Selective differentiation: redefining selections to appear in Acta Math. Acad. Sci. Hung.

[7] B. S. Thomson, Derivation bases on the real line, Real Anal. Exch. 8 (1982-83), 1, 67-207.

[8] _______, Differentiation bases on the real line, Real Anal. Exch. 8 (1982-83), 2, 278-442.

DEPARTMENT OF MATHEMATICS
UNIVERSITY OF CALIFORNIA-SANTA BARBARA
SANTA BARBARA, CA 93106

DEPARTMENT OF MATHEMATICS
UNIVERSITY OF WISCONSIN-MILWAUKEE
MILWAUKEE, WI 53201

DEPARTMENT OF MATHEMATICS
SIMON FRASER UNIVERSITY
BURNABY (2), BRITISH COLUMBIA
CANADA

Contemporary Mathematics
Volume 42, 1985

BAIRE ONE, NULL FUNCTIONS

A. M. Bruckner, J. Mařík, and C. E. Weil

ABSTRACT. It is proved that a Baire one function which is zero almost everywhere can be written as the product of two derivatives. Moreover, if the function is nonnegative, then the factors can be selected to be nonnegative. In both cases the factors can be chosen to have arbitrarily small L^p norm for $1 \leq p < \infty$.

1. INTRODUCTION. It has been known for some time now that the class of derivatives does not behave well with respect to multiplication. In fact since the turn of the century a number of authors have obtained results which indicate when the product of two derivatives is again a derivative. A relatively complete summary of such results is contained in Fleissner [2].

More recently the focus of attention has turned towards the question of determining the class of functions which can be expressed as the product of two or more derivatives. This work contains results along these lines. Specifically it is first shown that if φ is a bounded, Baire one function that is zero almost everywhere, then φ can be written as the product of two bounded derivatives. Next it is shown that even if φ is not bounded, it is still the product of two derivatives of arbitrarily small L^p norms, $1 \leq p < \infty$. In any case, if φ is nonnegative, then the factors can be selected to be nonnegative.

2. PRELIMINARIES. Let $R = (-\infty, \infty)$. The only measure used is Lebesgue measure in R and each integral should be interpreted as the corresponding Lebesgue integral. For each $S \subset R$, $|S|$ denotes its outer measure, and χ_S its characteristic function. All functions will be real valued functions of a real variable. If S is an open set or an interval, then $\Delta(S)$ is the family of all

---

1980 Mathematics Subject Classification. 26A21.

finitely differentiable functions on S where differentiability is one-sided in the case of an endpoint of S that belongs to S. Further set $\Delta^+(S) = \{F \in \Delta(S) : F' \geq 0 \text{ on } S\}$, $\mathcal{D}(S) = \{F' : F \in \Delta(S)\}$ and $\mathcal{D}^+(S) = \{f \in \mathcal{D}(S) : f \geq 0 \text{ on } S\}$. We write Δ, Δ^+ etc. for $\Delta(R)$, $\Delta^+(R)$ etc. Further let $\mathcal{J} = \{f \in \mathcal{D} : 0 \leq f < 2 \text{ on } R\}$.

The term "D-closed" refers to the Denjoy topology or the density topology on R as it is often called. (See for example [5]). Note that each set of measure zero is D-closed. Let C_{ap} denote the system of all functions approximately continuous on R (that is, continuous relative to the Denjoy topology). It is well known that each element of C_{ap} is a Baire one function.

Let $\mathfrak{A}$ be the collection of all sets $S \subset R$ such that S is both an F_σ set and a G_δ set. Such sets are often called ambiguous sets. As is well known, a function f is of Baire class one if and only if for each $c \in R$ $\{x : f(x) > c\}$ and $\{x : f(x) < c\}$ are F_σ sets. Consequently $S \in \mathfrak{A}$ if and only if χ_S is of Baire class one. It follows from Proposition 3 and Theorem 2 of [1] that $S \in \mathfrak{A}$ if and only if there are $F,G,H \in \Delta$ such that $\chi_S = F'G + H'$. The objective of the next section is to prove that there are $f,g \in \mathcal{J}$ such that $\chi_S = fg$ if and only if $S \in \mathfrak{A}$ and S is D-closed. This section is concluded with some facts that will be needed in the rest of the paper. The first has an easy proof which is therefore omitted. The second is a special case of the first.

2.1. THEOREM. Let V be open, A closed. Let $f \in \mathcal{D}(V \backslash A)$, $\alpha \in C_{ap}$. Let f and α be bounded and let $\alpha = 0$ on $V \cap A$. Set $f^* = 0$ on $V \cap A$, $f^* = \alpha f$ on $V \backslash A$. Then $f^* \in \mathcal{D}(V)$.

2.2. THEOREM. If f and α are bounded functions with $f \in \mathcal{D}$ and $\alpha \in C_{ap}$, then $\alpha f \in \mathcal{D}$.

The next result is a part of Theorem 3.2 in [5].

2.3. THEOREM. If $|S| = 0$ and if φ is a Baire one function on R, then there is an $\alpha \in C_{ap}$ such that $\alpha = \varphi$ on S.

Our next assertion follows from a theorem on p. 257 of [3] by letting $\alpha = 1$.

2.4. THEOREM. Let A be a G_δ set and B, an F_σ set. If $A \subset B$, then there is an $M \in \mathfrak{A}$ such that $A \subset M \subset B$.

We now restate Lemma 12 on p. 29 of [6].

2.5. THEOREM. Let A and B be disjoing, D-closed, G_δ sets. Then there is an $\alpha \in C_{ap}$ such that $0 \leq \alpha \leq 1$ on R, $\alpha = 1$ on A, and $\alpha = 0$ on B.

Finally we establish a consequence of the above which will prove useful in the rest of the work.

2.6. THEOREM. Let A and B be as in 2.5. Then there are $\alpha,\beta \in C_{ap}$ such that $0 \le \alpha \le 1$, $0 \le \beta \le 1$, $\alpha\beta = 0$ on R, $\alpha = 1$ on A and $\beta = 1$ on B.

PROOF. By 2.5 there is a $\gamma \in C_{ap}$ such that $\gamma = 1$ on A and $\gamma = 0$ on B. Let $A_1 = \{x : \gamma(x) \ge \frac{1}{2}\}$, $B_1 = \{x : \gamma(x) \le \frac{1}{2}\}$. Then A and B_1 are disjoint, D-closed, G_δ sets. Applying 2.5 again we get an $\alpha \in C_{ap}$ such that $0 \le \alpha \le 1$ on R, $\alpha = 1$ on A and $\alpha = 0$ on B_1. Similarly there is a $\beta \in C_{ap}$ such that $0 \le \beta \le 1$ on R, $\beta = 0$ on A_1 and $\beta = 1$ on B. Clearly $\alpha\beta = 0$ on R so that α and β satisfy our requirements.

3. D-CLOSED AMBIGUOUS SETS. As was mentioned above the objective of this section is to show that those sets whose characteristic functions are the products of two derivatives from $\mathcal{T}$ are precisely the D-closed ambiguous sets. We begin with a lemma which clearly establishes one direction of this characterization. It is actually an easy consequence of Theorem 5.5 of [4] but we include a proof here for the sake of completeness.

3.1. LEMMA. Let m be a natural number, $a,b \in R$, $a < b$; set $J = [a,b]$. Let $f_1,\dots,f_m \in \mathcal{B}^+(J)$, $f_1 \cdots f_m = \chi_S$ on J, $\limsup_{x \searrow 0} |S \cap (a,x)|/(x-a) > 0$. Then $a \in S$.

PROOF. There is an $\epsilon \in (0,\infty)$ and numbers $x_n \in (a,b)$ such that $x_n \to a$ and $|S \cap (a,x_n)| > \epsilon(x_n - a)$ for $n = 1,2,\dots$. Choose an n and set $L = [a,x_n]$. By Hölder's inequality we have $\epsilon \le \frac{1}{|L|}\int_L (f_1 \cdots f_m)^{1/m} \le \Pi_{j=1}^m \left(\frac{1}{|L|}\int_L f_j\right)^{1/m}$. This easily implies that $\Pi_{j=1}^m f_j(a) > 0$ so that $a \in S$.

For the remainder of this section let S denote a fixed subset of R. As a notational convenience for each interval $J \subset R$ we will let $\mathcal{P}(J)$ be the set of all pairs (f,g), where $f,g \in \mathcal{B}^+(J)$, $f < 2$, $g < 2$ on J, $f = g = 1$ on $J \cap S$ and $fg = 0$ on $J \setminus S$. Let $\mathcal{J}$ be the system of all intervals J such that $\mathcal{P}(J) \ne \emptyset$.

We now prove four lemmas followed by the other direction of the desired characterization.

3.2. LEMMA. Let $a_1,a_2,b_1,b_2 \in R$, $a_1 < a_2 < b_1 < b_2$. Let $(f_j,g_j) \in \mathcal{P}([a_j,b_j])$ $(j = 1,2)$. Then there is a pair $(f,g) \in \mathcal{P}([a_1,b_2])$ such that $f = f_1$, $g = g_1$ on $[a_1,a_2]$ and $f = f_2$, $g = g_2$ on $[b_1,b_2]$.

PROOF. If there is a $c \in [a_2,b_1] \cap S$, then $f_1(c) = \dots = g_2(c) = 1$ and we set $f = f_1$, $g = g_1$ on $[a_1,c]$, $f = f_2$, $g = g_2$ on $[c,b_2]$. Otherwise we have $f_1g_1 = f_2g_2 = 0$ on $[a_2,b_1]$.

Then we choose a $c \in (a_2,b_1)$ and construct functions f,g such that $f=f_1$, $g=g_1$ on $[a_1,a_2]$, $f(c)=g(c)=0$, $f=f_2$, $g=g_2$ on $[b_1,b_2]$ and that f and g are linear on each of the intervals $[a_2,c]$ and $[c,b_1]$. It is easy to see that $(f,g) \in \mathcal{P}(]a_1,b_2])$.

3.3. LEMMA. Let L be an open interval. Suppose that for each $x \in L$ there is an open interval I such that $x \in I \in \mathcal{J}$. Then $L \in \mathcal{J}$.

PROOF. Choose numbers $x_n \in L$ $(n=0,\pm 1,\pm 2,\dots)$ such that

(1) $$x_{n-1} < x_n, \quad \inf_n x_n = \inf L, \quad \sup_n x_n = \sup L .$$

It follows easily from 3.2 that $[x_{n-1},x_{n+1}] \in \mathcal{J}$ for each n. Let $(f_n,g_n) \in \mathcal{P}([x_{n-1},x_{n+1}])$. Applying 3.2 once more we get a pair $(f,g) \in \mathcal{P}(L)$ such that $f(x_n)=f_n(x_n)$, $g(x_n)=g_n(x_n)$ for each n.

3.4. LEMMA. Let $a,b \in R$, $a<b$, $q \in (0,b-a)$. Set $J=[a,b]$. Let $(f_0,g_0) \in \mathcal{P}(J)$. Then there is a pair $(f,g) \in \mathcal{P}(J)$ such that $f=f_0$, $g=g_0$ on $[a,b]$, $q<\int_J f \le |J|$ and $q<\int_J g \le |J|$.

PROOF. Choose a $p \in (0,1)$ such $|J|p > q(2-p)$. Set $S_1=(S\cap J)\cup\{a,b\}$, $S_0=J\setminus S_1$. It follows from 3.1 that S_1 is a D-closed, G_δ set. If $|S_0|=0$, we define $A=B=\emptyset$. Otherwise $|S_0|/(2-p) < |S_0|$ and we choose disjoint closed sets $A,B \subset S_0$ such that $|A|=|B|$ and that $|A\cup B| > |S_0|/(2-p)$. It follows from 2.6 that there are $\alpha,\beta,\gamma \in C_{ap}$ with $0 \le \alpha \le 1$, $0\le\beta\le 1$, $0\le\gamma\le 1$, $\alpha\beta=\alpha\gamma=\beta\gamma=0$ on R, $\alpha=1$ on A, $\beta=1$ on B and $\gamma=1$ on S_1. Define $f=2p\alpha+\gamma f_0$, $g=2p\beta+\gamma g_0$ on J. It is easy to see that $(f,g)\in\mathcal{P}(J)$ and that $f=f_0$, $g=g_0$ on $\{a,b\}$. Since $2|A|(2-p) \ge |S_0|$, we have $q<\frac{p}{2-p}(|S_0|+|S_1|) \le |S_1|+2p|A| \le \int_J f \le$
$\le |S_1|+2p|A|+2(|S_0|-2|A|) = |J|+|S_0|-2|A|(2-p) \le |J|$; similarly for g.

3.5. LEMMA. Let $a,b \in R$, $a<b$; set $L=(a,b)$. Let $L \in \mathcal{J}$ and let w be a positive, continuous function on L. Then there is a pair $(f,g)\in\mathcal{P}(L)$ such that for each $x \in L$

(2) $$\max\left(\left|\int_a^x (f-1)\right|, \left|\int_a^x (g-1)\right|\right) \le \int_a^x w ,$$
$$\max\left(\left|\int_x^b (f-1)\right|, \left|\int_x^b (g-1)\right|\right) \le \int_x^b w .$$

PROOF. Set $\psi=w/2$. We may suppose that $\psi<1$ on L. There are numbers $x_n \in L$ $(n=0,\pm1,\pm2,\dots)$ fulfilling (1) such that

$$x_n - x_{n-1} < \min\left(\int_a^{x_{n-1}} \psi, \int_{x_n}^b \psi\right) \text{ for each } n .$$

Let $(f^*,g^*) \in \mathcal{P}(L)$. For each n set $J_n = [x_{n-1}, x_n]$. By 3.4 there are pairs $(f_n, g_n) \in \mathcal{P}(J_n)$ such that $f_n = f^*$, $g_n = g^*$ on $\{x_{n-1}, x_n\}$ and that

$$\int_{J_n} (1-\psi) < \int_{J_n} f_n \le |J_n| , \qquad \int_{J_n} (1-\psi) < \int_{J_n} g_n \le |J_n| .$$

Define functions f and g on L setting $f = f$, $g = g_n$ on J_n. Now let $x \in J_n$. Then $\int_a^x f \le \int_a^{x_n} f \le x_n - a < \int_a^{x_{n-1}} (1+\psi) < \int_a^x (1+w)$, $\int_a^x f \ge \int_a^{x_{n-1}} f > \int_a^{x_{n-1}} (1-\psi) > \int_a^{x_n} (1-\psi) - (x_n - x_{n-1}) > \int_a^{x_n} (1-w) \ge \int_a^x (1-w)$. Similarly for $\int_x^b f$ and for g. This proves (2). It is obvious that $(f,g) \in \mathcal{P}(L)$.

3.6. THEOREM. Let S be a D-closed ambiguous set. Then $R \in \mathcal{J}$.

PROOF. Let U be the set of all points x such that $x \in I$ for some open interval $I \in \mathcal{J}$. Let $A = R \setminus U$. Then A is closed. Suppose that $A \neq \emptyset$. Let w be a continuous function on R such that $w = 0$ on A and $w > 0$ on U. Since χ_S is a Baire one function, there is a bounded open interval I such that $A \cap I \neq \emptyset$ and that χ_S is constant on $A \cap I$. For each component $L = (a,b)$ of $I \cap U$ we have, by 3.3, $L \in \mathcal{J}$ so that by 3.5 there is a pair $(f,g) \in \mathcal{P}(L)$ fulfilling (2). In this way we construct functions f, g on $I \cap U$. Now we distinguish two cases.

Suppose first that $A \cap I \subset S$. Define $f^* = f$, $g^* = g$ on $I \cap U$, $f^* = g^* = 1$ on $I \cap A$. If $x_1, x_2 \in I$, $x_1 < x_2$ and if $x_1 \in A$ or $x_2 \in A$, then it follows easily from (2) that $|\int_{x_1}^{x_2} (f^*-1)| \le \int_{x_1}^{x_2} w$. Since w is continuous and $w = 0$ on A, $f^* \in \mathcal{D}(I)$; similarly $g^* \in \mathcal{D}(I)$. It is obvious that $(f^*,g^*) \in \mathcal{P}(I)$ so that $I \in \mathcal{J}$, $I \subset U$ – a contradiction.

Now suppose that $A \cap I \cap S = \emptyset$. Let J be a closed interval with interior V such that $A \cap V \neq \emptyset$ and $J \subset I$. Then $A \cap J$ and S are disjoint, D-closed, G_δ sets. By 2.5 there is an $\alpha \in C_{ap}$ such that $\alpha = 0$ on $A \cap J$, $\alpha = 1$ on S and $0 \le \alpha \le 1$ on R. Define $f^* = g^* = 0$ on $V \cap A$, $f^* = \alpha f$, $g^* = \alpha g$ on $V \cap U$. By 2.1 we have f^*, $g^* \in \mathcal{D}(V)$. It is obvious that $(f^*,g^*) \in \mathcal{P}(V)$ so that $V \in \mathcal{J}$, $V \subset U$ – a contradiction.

It follows that $A = \emptyset$. By 3.3 we have $R = U \in \mathcal{J}$.

3.7. COROLLARY. Let $S \subset R$. Then the following four conditions are equivalent:

1) There is a natural number m and functions $f_1, \dots, f_m \in \mathcal{D}^+$ such that $f_1 \cdots f_m = \chi_S$.

2) S is ambiguous and D-closed.

3) There are $f,g \in \mathcal{D}^+$ such that $f = g = 1$ on S, $fg = 0$ on $R \setminus S$ and $f < 2$, $g < 2$ on R.

4) There are $f,g \in \mathcal{T}$ such that $fg = \chi_S$.

PROOF. If 1) holds, then S is ambiguous and it follows easily from 3.1 that S is D-closed. If 2) holds, then, by 3.6, 3) holds as well. The implications 3) => 4) and 4) => 1) are obvious.

4. BOUNDED, BAIRE ONE, NULL FUNCTIONS. The goal of this section is to establish that a bounded, [nonnegative] Baire one function that is zero a.e. can be expressed as the product of two bounded [nonnegative] derivatives. This fact follows easily from Theorem 4.2 whose proof relies mainly on Theorem 3.6. We begin with a proposition based on 2.5 which is used in the proof of 4.2.

4.1. PROPOSITION. Let $A_1, A_2, \ldots$ be pairwise disjoint elements of $\mathfrak{A}$ of measure zero. Then there are $\alpha_1, \alpha_2, \ldots \in C_{ap}$ such that for each j $\alpha_j = 1$ on A_j, $0 \le \alpha_j \le 1$ on R and $\alpha_i \alpha_j = 0$ on R, if $i \ne j$.

PROOF. For each j let $S_j = \bigcup_{i \ne j} A_i$. Then $|S_j| = 0$. Let T_j be a G_δ set such that $S_j \subset T_j$ and $|T_j| = 0$. Further let $B_j = T_j \setminus A_j$. Then $S_j \subset B_j$; B_j and A_j are disjoint, D-closed G_δ sets. So by 2.5 there are $\alpha_j^*, \beta_j \in C_{ap}$ such that for each j $\alpha_j^* = 1$ on A_j, $\beta_j = 1$ on B_j, and $0 \le \alpha_j^* \le 1$, $0 \le \beta_j \le 1$, $\alpha_j^* \beta_j = 0$ on R. Let $\alpha_j = \beta_1 \cdots \beta_{j-1} \alpha_j^*$. If $i < j$, then $\alpha_i \alpha_j$ is a multiple of $\alpha_i^* \beta_i$ so that $\alpha_i \alpha_j = 0$. The other requirements are easily verified.

4.2. THEOREM. Let $B_1, B_2, \ldots$ be pairwise disjoint elements of $\mathfrak{A}$ of measure zero and let $Q \in (2, \infty)$. Let $\varphi_1, \varphi_2, \ldots$ be Baire one functions such that $|\varphi_n| \le 1$ for each n and that $\varphi_n \to 0$ uniformly on R. Then there are $f,g \in \mathcal{D}$ such that $|f| < Q$, $|g| < Q$ and $fg = \Sigma_{n=1}^{\infty} \varphi_n \chi_{B_n}$ on R. If, moreover, $\varphi_n \ge 0$ for each n, we may choose $f \ge 0$ and $g \ge 0$.

PROOF. Let $\epsilon_1, \epsilon_2, \ldots$ be positive numbers such that $\epsilon_1 = 1$ and that $\Sigma_{j=1}^{\infty} 2\epsilon_j < Q$. There are integers r_j such that $0 = r_0 < r_1 < \ldots$ and that $|\varphi_n| \le \epsilon_j^2$ for each $n > r_{j-1}$ $(j = 1,2,\ldots)$. Set $S_j = \{r_{j-1} + 1, \ldots, r_j\}$, $\psi_j = \Sigma_{n \in S_j} \varphi_n \chi_{B_n}$, and $A_j = \bigcup_{n \in S_j} B_n$. Then $A_1, A_2, \ldots$ are pairwise disjoint elements of $\mathfrak{A}$ of measure zero. Let α_j be as in 4.1. According to 2.3 there are $\gamma_1, \gamma_2, \ldots \in C_{ap}$ such that $\gamma_j = \psi_j$ on A_j. Since $|\psi_j| \le \epsilon_j^2$, we may assume $|\gamma_j| \leq \epsilon_j^2$. According to 3.6 there are $f_j, g_j \in \mathcal{T}$ such that $f_j g_j = \chi_{A_j}$. Obviously $|\alpha_j \gamma_j f_j/\epsilon_j| \leq 2\epsilon_j$ and $|\alpha_j g_j \epsilon_j| \leq 2\epsilon_j$. Define $f = \Sigma_{j=1}^{\infty} \alpha_j \gamma_j f_j/\epsilon_j$ and $g = \Sigma_{j=1}^{\infty} \alpha_j g_j \epsilon_j$. Then $|f| \leq \Sigma_{j=1}^{\infty} 2\epsilon_j < Q$ and also $|g| < Q$. The series that define f and g converge uniformly and by 2.2 each term is in $\mathcal{D}$. Thus $f, g \in \mathcal{D}$. Since $A_1, A_2, \ldots$ are pairwise disjoint and $\gamma_j \chi_{A_j} = \psi_j$, we have $fg = \Sigma_{j=1}^{\infty} \alpha_j^2 \gamma_j f_j g_j = \Sigma_{j=1}^{\infty} \alpha_j^2 \psi_j = \Sigma_{k=1}^{\infty} \psi_j = \Sigma_{n=1}^{\infty} \varphi_n \chi_{B_n}$. If $\varphi_n \geq 0$ for each n, then we may choose $\gamma_j \geq 0$ and we obtain $f \geq 0$ and $g \geq 0$ on R.

4.3. COROLLARY. Let φ be a Baire one function, $|\varphi| \leq 1$ on R, $\varphi = 0$ a.e. and let $Q \in (2, \infty)$. Then there are $f, g \in \mathcal{D}$ such that $|f| < Q$, $|g| < Q$ and $fg = \varphi$ on R. Moreover, if $\varphi \geq 0$, we may select $f \geq 0$ and $g \geq 0$.

PROOF. Let a_n be numbers such that $a_0 > a_1 > \ldots$, $a_1 = 1$ and $a_n \to 0$. For $n = 1, 2, \ldots$ let $V_n = \{x : a_{n+1} \leq |\varphi(x)| \leq a_n\}$ and $W_n = \{x : 0 < |\varphi(x)| < a_{n-1}\}$. Then $V_n \subset W_n$, V_n is a G_δ set and W_n an F_σ set. So by 2.4 there are $M_n \in \mathfrak{A}$ such that $V_n \subset M_n \subset W_n$. Let $B_n = M_n \setminus \bigcup_{j=1}^{n-1} M_j$ for $n = 1, 2, \ldots$. We see that $B_1, B_2, \ldots$ are pairwise disjoint elements of $\mathfrak{A}$ of measure zero, $\bigcup_{n=1}^{\infty} B_n = \bigcup_{n=1}^{\infty} M_n = \{x : \varphi(x) \neq 0\}$, and $\varphi = \Sigma_{n=1}^{\infty} \varphi \chi_{B_n}$. Since $B_n \subset W_n$, we have $\varphi \chi_{B_n} \to 0$ uniformly on R. Now we apply 4.2.

The next result shows that we would get a wrong assertion, if we admitted $Q < 2$ in 4.3. If, however, the function φ in 4.3 satisfies the relation $\varphi(R) = \{0, 1\}$, then, by 3.6, it can be expressed as the product of two nonnegative derivatives each of which is bounded by 2. We do not know whether Q can be replaced by 2 in 4.3.

4.4. THEOREM. Let $Q \in (0, \infty)$. Let $f, g \in \mathcal{D}$, $|f| \leq Q$, $|g| \leq Q$ and $fg = 0$ a.e. on R. Then $|fg| \leq Q^2/4$ on R.

PROOF. Let $x, y \in R$, $y \neq x$. Let $a = (y-x)^{-1} \int_x^y f$, $b = (y-x)^{-1} \int_x^y g$. Since $|f| + |g| \leq Q$ a.e., we have $|a| + |b| \leq Q$ so that $4|ab| \leq Q^2$. Thus $|f(x)g(x)| \leq Q^2/4$.

5. ARBITRARY BAIRE ONE, NULL FUNCTIONS. Throughout the rest of the paper p will denote a number in $[1, \infty)$. We set $\gamma = 1 - p^{-1}$. If f is a function, $S \subset R$ and if the integral $M = \int_S |f|^p$ is finite, we write $\|f\|_S = M^{1/p}$. If, moreover, $|S| < \infty$, then, by Hölder's inequality,

(3) $$\int_S |f| \leq \|f\|_S \cdot |S|^\gamma .$$

If the meaning of S is obvious from the context (if, e.g., S is the domain of definition of f), we write $\|f\|_S = \|f\|$. The class of all functions φ for which to each $\epsilon \in (0,\infty)$ there correspond $f,g \in \mathcal{D}$ ($\mathcal{D}^+$ resp.) such that $\|f\|_R + \|g\|_R < \epsilon$ and $fg = \varphi$ on S will be denoted by $\mathcal{R}(S)$ ($\mathcal{R}^+(S)$ resp.). Moreover, $\ell oc\,\mathcal{R}(S)$ ($\ell oc\,\mathcal{R}^+(S)$ resp.) is the family of all functions φ such that for each $x \in S$ there is an open interval I with $x \in I$ and $\varphi \in \mathcal{R}(S \cap I)$ ($\mathcal{R}^+(S \cap I)$ resp.). In keeping with our previous conventions $\mathcal{R}(R)$ and $\mathcal{R}^+(R)$ will be denoted simply by $\mathcal{R}$ and $\mathcal{R}^+$ respectively.

In terms of this notation our objective is to show that if φ is a Baire one function which is zero a.e. on R, then $\varphi \in \mathcal{R}$. If in addition $\varphi \geq 0$, then $\varphi \in \mathcal{R}^+$. We begin with a useful lemma and the treat the case where φ is bounded.

5.1. LEMMA. Let A be a measurable set and let $\epsilon \in (0,\infty)$. Then there is a closed set $C \subset A$ and a $\lambda \in C_{ap}$ such that $|A \setminus C| < \epsilon$, $0 \leq \lambda \leq 1$ on R, $\lambda = 1$ on C and $\lambda = 0$ on $R \setminus A$.

PROOF. Denote by S the set of all points in A that are points of density of A. There is an F_σ set $T \subset S$ such that $|S \setminus T| = 0$. There is a closed set $C \subset T$ such that $|T \setminus C| < \epsilon$. Clearly $|A \setminus C| = |T \setminus C| < \epsilon$. Since $R \setminus T$ and C are disjoint, D-closed, G_δ sets, by 2.5 there is a $\lambda \in C_{ap}$ such that $\lambda = 1$ on C, $\lambda = 0$ on $R \setminus T$, and $0 \leq \lambda \leq 1$ on R.

5.2. THEOREM. Let φ be a bounded Baire one function such that $\varphi = 0$ a.e. on R. Then $\varphi \in \mathcal{R}$. If in addition $\varphi \geq 0$, then $\varphi \in \mathcal{R}^+$.

PROOF. Assume as we may that $|\varphi| \leq 1$. Let $\epsilon \in (0,\infty)$. By 4.3 there are $f_1, g_1 \in \mathcal{D}$ such that $|f_1| < 3$, $|g_1| < 3$, and $f_1 g_1 = \varphi$ on R. Let $B = \{x : \varphi(x) \neq 0\}$. It follows easily from 5.1 that there is an open set $U \supset B$ and a $\lambda \in C_{ap}$ such that $|U| = |U \setminus B| < (\epsilon/6)^p$, $0 \leq \lambda \leq 1$ on R, $\lambda = 1$ on B and $\lambda = 0$ on $R \setminus U$. Let $f = \lambda f_1$ and $g = \lambda g_1$. By 2.2 $f,g \in \mathcal{D}$. Since $\lambda = 1$ on B, $fg = f_1 g_1 = \varphi$. Finally $\|f\| + \|g\| \leq 2 \cdot 3 \cdot |U|^{1/p} < \epsilon$. If $\varphi \geq 0$, then we may select f_1 and g_1 from $\mathcal{D}^+$ and we have $f,g \in \mathcal{D}^+$.

We now take up the process of showing that the assumption of boundedness can be deleted from 5.2. We start with three assertions the first of which will be used again later.

5.3. LEMMA. Let $J = [a,b]$ and let $\epsilon \in (0,\infty)$. Let $f_1, g_2, g_1, g_2 \in \mathcal{D}(J)$ ($\mathcal{D}^+(J)$ resp.). Suppose that $f_1 g_1 = f_2 g_2 = \varphi$ on J and that $\varphi = 0$ on a dense subset of J. Then there are a

$c \in (a,b)$ and $f,g \in \mathcal{B}(J)$ ($\mathcal{B}^+(J)$ resp.) such that $fg = \varphi$ on J, $f(c) = g(c) = 0$, $f(a) = f_1(a)$, $g(a) = g_1(a)$, $f(b) = f_2(b)$, $g(b) = g_2(b)$, $|f| \le |f_1| + \varepsilon$, $|g| \le |g_1| + \varepsilon$ on $[a,c]$ and $|f| \le |f_2| + \varepsilon$, $|g| \le |g_2| + \varepsilon$ on $[c,b]$.

PROOF. There is a $c \in (a,b)$ such that f_1, f_2, g_1, and g_2 are all continuous at c. Since $f_1 g_1 = 0$ on a dense subset of J, $f_1(c) g_1(c) = 0$. Since the roles of f_1 and g_1 are interchangeable, we may assume that $f_1(c) = 0$. There is a $c_1 \in (a,c)$ and a function h continuous on $[a,c]$ such that $h = 1$ on $[a,c_1]$, $h(c) = 0$, $|f_1|/\varepsilon < h \le 1$ on $[c_1,c)$ and g_1 is bounded on $[c_1,c]$. Set $f = f_1/\sqrt{h}$, $g = g_1\sqrt{h}$ on $[a,c)$ and $f(c) = g(c) = 0$. We define f and g on $[c,b]$ in a similar fashion. On $[c_1,c)$ we have $|f| < \varepsilon\sqrt{h} \le \varepsilon$; in particular, f is continuous at c. Now it is easy to see that $f \in \mathcal{B}(J)$. The rest of the proof is left to the reader.

5.4. PROPOSITION. Let $-\infty < a_1 < a_2 < b_1 < b_2 < \infty$ and let $\varphi \in \mathcal{R}([a_1,b_1]) \cap \mathcal{R}([a_2,b_2])$ ($\mathcal{R}^+([a_1,b_1]) \cap \mathcal{R}^+([a_2,b_2])$ resp.). Then $\varphi \in \mathcal{R}([a_1,b_2])$ ($\mathcal{R}^+([a_1,b_2])$ resp.).

PROOF. Let $\varepsilon \in (0,\infty)$. There are $f_1, g_1, f_2, g_2 \in \mathcal{B}$ ($\mathcal{B}^+$ resp.) such that $f_1 g_1 = \varphi$ on $[a_1,b_1]$, $f_2 g_2 = \varphi$ on $[a_2,b_2]$ and $\|f_1\|^p + \|g_1\|^p + \|f_2\|^p + \|g_2\|^p < (\varepsilon/2)^p$. Let $J = [a_2,b_1]$. It follows easily from 5.3 that there are $f,g \in \mathcal{B}(J)$ ($\mathcal{B}^+(J)$ resp.) such that $fg = \varphi$ on J, $f(a_2) = f_1(a_2)$, $g(a_2) = g_1(a_2)$, $f(b_1) = f_2(b_1)$, $g(b_1) = g_2(b_1)$ and $\|f\|_J^p + \|g\|_J^p < (\varepsilon/2)^p$. Let $f = f_1$, $g = g_1$ on $(-\infty,a_2)$ and $f = f_2$, $g = g_2$ on (b_1,∞). Then $f,g \in \mathcal{B}$ ($\mathcal{B}^+$ resp.), $fg = \varphi$ on $[a_1,b_2]$ and $\|f\|^p + \|g\|^p < 2(\varepsilon/2)^p$. So $(\|f\| + \|g\|)^p \le 2^{p-1}(\|f\|^p + \|g\|^p) < \varepsilon^p$.

The next statement follows easily from the above by a routine compactness argument.

5.5. COROLLARY. Let J be a closed, bounded interval and let $\varphi \in \ell oc\,\mathcal{R}(J)$ ($\ell oc\,\mathcal{R}^+(J)$ resp.). Then $\varphi \in \mathcal{R}(J)$ ($\mathcal{R}^+(J)$ resp.).

The preceding result and the next lemma are used in the proofs of propositions 5.7 and 5.7.1.

5.6. LEMMA. Let $J = [a,b]$. Let $M,N \in R$, $f_1, g_1 \in \mathcal{B}(J)$, $f_1 g_1 = 0$ a.e. on J and $\|f_1\| + \|g_1\| < \infty$. Then there are $f,g \in \mathcal{B}(J)$ such that $f = f_1$, $g = g_1$ on $\{a,b\}$, $fg = f_1 g_1$ on J, $\int_J f = M$, $\int_J g = N$, $\|f\| \le 5\|f_1\| + |M|(4/|J|)^\gamma$, $\|g\| \le \|g_1\| + |N|(4/|J|)^\gamma$. If in addition $f_1, g_1 \in \mathcal{B}^+(J)$, $M \ge \int_J f_1$ and $N \ge \int_J g_1$, then f and g can be chosen from $\mathcal{B}^+(J)$.

PROOF. Let $B = \{x : f_1(x)g_1(x) \neq 0\} \cup \{x : f_1$ or g_1 is not approximately continuous at $x\} \cup \{a,b\}$. Then $|B| = 0$. There is a $K \in (0,\infty)$ such that $|S| > |J|/2$ where $S = \{x \in J\setminus B : |f_1(x)| + |g_1(x)| \leq K\}$. There is a closed set $A \subset S$ with $|A| > |J|/2$. According to 5.1 there is a closed set $C \subset A$ and a $\lambda \in C_{ap}$ such that $|C| > |J|/2$, $0 \leq \lambda \leq 1$ on R, $\lambda = 1$ on C and $\lambda = 0$ on $R\setminus A$. Note that λf_1 is approximately continuous on A (since $A \subset J\setminus B$) and $\lambda f_1 = 0$ on $J\setminus A$ which is open in J. It follows that λf_1 is approximately continuous on J. Since $|\lambda f_1| \leq K$ on J, $\lambda f_1 \in \mathcal{B}(J)$. Likewise $\lambda g_1 \in \mathcal{B}(J)$. Let $f_2 = f_1 - \lambda f_1$ and $g_2 = g_1 - \lambda g_1$. Then $f_2, g_2 \in \mathcal{B}(J)$ and, as is easily verified from the properties of λ, $f_2 g_2 = f_1 g_1$, $|f_2| \leq |f_1|$ and $|g_2| \leq |g_1|$. Since $\|f_1\| < \infty$, we have $\int_J |f_2| \leq \int_J |f_1| < \infty$; similarly $\int_J |g_2| < \infty$. It follows from 5.1 that there are $\alpha, \beta \in C_{ap}$ and disjoint, closed subsets C_α and C_β of C such that $|C_\alpha| > |J|/4$, $|C_\beta| > |J|/4$, $0 \leq \alpha \leq 1$, $0 \leq \beta \leq 1$, $\alpha\beta = 0$ on R, $\alpha = \beta = 0$ on $R\setminus C$, $\alpha = 1$ on C_α and $\beta = 1$ on C_β. Let $s = (M - \int_J f_2)/\int_J \alpha$ and $t = (N - \int_J g_2)/\int_J \beta$. Let $f = f_2 + s\alpha$ and $g = g_2 + t\beta$. Obviously $\alpha^p \leq \alpha$ and $\int_J \alpha \geq |J|/4$; by (3), $\int_J |f_2| \leq \|f_1\| \cdot |J|^\gamma$. Thus $(\int_J \alpha^p)^{1/p}/\int_J \alpha \leq (\int_J \alpha)^{p^{-1}-1} \leq (4/|J|)^\gamma$, $\|f\| \leq \|f_2\| + \|\alpha\|(|M| + \int_J |f_2|)/\int_J \alpha \leq \|f_1\| + (4/|J|)^\gamma(|M| + \|f_1\| \cdot |J|^\gamma) \leq 5\|f_1\| + (4/|J|)^\gamma |M|$. A similar estimate is valid for g.

If the additional assumption is fulfilled, then $f_2 \geq 0$, $g_2 \geq 0$, $s \geq 0$, $t \geq 0$ so that $f \geq 0$ and $g \geq 0$.

5.7. PROPOSITION. Let I be an open interval, ω a positive, continuous function on I, $F_0, G_0 \in \Delta(I)$, $\|F_0'\| + \|G_0'\| < \infty$, $\varphi \in \ell oc\ (I)$ and let $\epsilon \in (0,\infty)$. Then there are $F, G \in \Delta(I)$ such that

(4) $\quad F'G' = \varphi$, $|F - F_0| + |G - G_0| < \omega$ on I and

(5) $\quad \|F'\|^p + \|G'\|^p < \epsilon + 8^{p-1}(\|F_0'\|^p + \|G_0'\|^p)$.

PROOF. There are numbers $y_n \in I$ $(n = 0, \pm 1, \pm 2, \ldots)$ such that $y_n < y_{n+1} < y_n + \frac{1}{2}$, $\inf_n y_n = \inf I$, $\sup_n y_n = \sup I$ and that

$$\int_{y_{n-1}}^{y_{n-1}} (|F_0'| + |G_0'|)| < \mu_n = \min\{\omega(x) : x \in [y_{n-1}, y_{n+1}]\}/7 .$$

Choose $\epsilon_n \in (0, \mu_n^p)$ such that $\Sigma_{n=-\infty}^{\infty} \epsilon_n < \epsilon/2^{p-1}$. Since $\varphi \in \ell oc\,\mathcal{R}(I)$, 5.5 implies that $\varphi \in \mathcal{R}([y_{n-1}, y_{n+1}])$. There are $f_{n,1}, g_{n,1} \in \mathcal{B}$ such that $f_{n,1} g_{n,1} = \varphi$ on $[y_{n-1}, y_{n+1}]$ and $\|f_{n,1}\|^p + \|g_{n,1}\|^p < \epsilon_n/5^p$. It follows from 5.3 that there are $x_n \in (y_{n-1}, y_n)$ and $f_{n,2}, g_{n,2} \in \mathcal{B}([x_n, x_{n+1}])$ such that $f_{n,2} g_{n,2} = \varphi$ on $[x_n, x_{n+1}]$, $f_{n,2} = g_{n,2} = 0$ on $\{x_n, x_{n+1}\}$, and $\|f_{n,2}\|^p + \|g_{n,2}\|^p < \epsilon_n/5^p$. Denote

$[x_n, x_{n+1}]$ by J_n. According to 5.6 there are $f_n, g_n \in \mathcal{D}(J_n)$ such that $f_n = g_n = 0$ on $\{x_n, x_{n+1}\}$, $f_n g_n = \varphi$ on J_n,

(6) $$\int_{J_n} f_n = \int_{J_n} F_0', \quad \int_{J_n} g_n = \int_{J_n} G_0',$$

(7) $$\|f_n\| \leq 5\|f_{n,2}\| + |\int_{J_n} F_0'|(4/|J_n|)^\gamma,$$

(8) $$\|g_n\| \leq 5\|g_{n,2}\| + |\int_{J_n} G_0'|(4/|J_n|)^\gamma.$$

Using (3), (7), and the relation $|J_n| < 1$ we get

$\int_{J_n} |f_n| \leq 5|J_n|^\gamma \|f_{n,2}\| + 4^\gamma \int_{J_n} |F_0'| \leq \epsilon_n^{1/p} + 4\int_{J_n} |F_0'|$,

$\|f_n\|^p \leq 2^{p-1}(5^p\|f_{n,2}\|^p + 4^{p\gamma}\|F_0'\|^p_{J_n})$. Similarly,

$\int_{J_n} |g_n| \leq \epsilon_n^{1/p} + 4\int_{J_n} |G_0'|$, $\|g_n\|^p \leq 2^{p-1}(5^p\|g_{n,2}\|^p + 4^{p\gamma}\|G_0'\|^p_{J_n})$. There are $F, G \in \mathcal{D}(I)$ such that $F' = f_n$, $G' = g_n$ on J_n, $F(x_n) = F_0(x_n)$, and $G(x_n) = G_0(x_n)$ for each n. If $x \in J_n$, then (since $\epsilon_n < \mu_n^p$ and $J_n \subset [y_{n-1}, y_{n+1}]$) $|F(x) - F_0(x)| + |G(x) - G_0(x)| \leq \int_{J_n} (|f_n| + |g_n| + |F_0'| + |G_0'|) < 2\epsilon_n^{1/p} + 4\mu_n + \mu_n \leq \omega(x)$. Finally (note that $p\gamma = p-1$)

$\|F'\|^p + \|G'\|^p = \Sigma_{n=-\infty}^{\infty}(\|f_n\|^p + \|g_n\|^p) \leq 2^{p-1}\Sigma_{n=-\infty}^{\infty}\epsilon_n +$

$8^{p-1}\Sigma_{n=-\infty}^{\infty}(\|F_0'\|^p_{J_n} + \|G_0'\|^p_{J_n}) < \epsilon + 8^{p-1}(\|F_0'\|^p_I + \|G_0'\|^p_I)$.

The version of the preceding theorem involving nonnegative functions is somewhat different. Consequently we state it separately.

5.7.1. PROPOSITION. Let I be an open interval, ω a positive, continuous function on I, $F_0, G_0 \in \Delta^+(I)$, $F_0'G_0' > 0$ on I, $\|F_0'\| + \|G_0'\| < \infty$, $\varphi \in loc\,\mathcal{R}^+(I)$ and let $\epsilon \in (0,\infty)$. Then there are $F, G \in \Delta^+(I)$ fulfilling (4) and (5).

PROOF. There are $x_n, y_n \in I$ such that $x_n < y_n < x_{n+1} < x_n + \frac{1}{2}$ $(n = 0, \pm 1, \pm 2, \ldots)$, $\inf_n x_n = \inf I$, $\sup_n x_n = \sup I$ and $F_0(x_{n+1}) - F_0(y_{n-1}) + G_0(x_{n+1}) - G_0(y_{n-1}) < \min\{\omega(x) : x \in [y_{n-1}, x_{n+1}]\}$ for each n. Choose $\epsilon_n \in (0,\infty)$ such that $\Sigma_{n=-\infty}^{\infty}\epsilon_n < \epsilon/2^{p-1}$. Since $\varphi \in loc\,\mathcal{R}^+(I)$, 5.5 implies that $\varphi \in \mathcal{R}^+([y_{n-1}, x_{n+1}])$. Consequently there are $f_{n,1}, g_{n,1} \in \mathcal{D}^+$ such that $f_{n,1}g_{n,1} = \varphi$ on $[y_{n-1}, x_{n+1}]$ and $\|f_{n,1}\|^p + \|g_{n,1}\|^p < \zeta_n = \min\{\epsilon_n/5^p, (F_0(y_n) - F_0(x_n))^p, (G_0(y_n) - G_0(x_n))^p\}$. It follows from 5.3 that there are $z_n \in (y_n, x_{n+1})$ and functions $f_{n,2}, g_{n,2} \in \mathcal{D}^+([z_{n-1}, z_n])$ such that $f_{n,2}g_{n,2} = \varphi$ on $[z_{n-1}, z_n]$, $f_{n,2} = g_{n,2} = 0$ on $\{z_{n-1}, z_n\}$, and $\|f_{n,2}\|^p + \|g_{n,2}\|^p < \zeta_n$. Denote $[z_{n-1}, z_n]$ by J_n. Since $|J_n| < 1$, we have (see (3)) $\int_{J_n} f_{n,2} \leq$

$\|f_{n,2}\| < F_0(y_n) - F_0(x_n) < F_0(z_n) - F_0(z_{n-1})$; similarly $\int_{J_n} g_{n,2} < G_0(z_n) - G_0(z_{n-1})$. According to 5.6 there are $f_n, g_n \in \mathcal{D}^+(J_n)$ fulfilling (6)-(8) such that $f_n = g_n = 0$ on $\{z_{n-1}, z_n\}$ and $f_n g_n = \varphi$ on J_n. There are $F, G \in \mathcal{D}^+(I)$ such that $F' = f_n$, $G' = g_n$ on J_n, $F(z_n) = F_0(z_n)$ and $G(z_n) = G_0(z_n)$ for each n. The estimate (5) follows just as it did in the proof of 5.7. It is clear that $F, G \in \Delta^+(I)$ and that $F'G' = \varphi$ on I. If $x \in J_n$, then $F_0(z_{n-1}) \le F(x) \le F_0(z_n)$ so that $|F(x) - F_0(x)| \le F_0(z_n) - F_0(z_{n-1}) < F_0(x_{n+1}) - F_0(y_{n-1})$. Likdwise $|G(x) - G_0(x)| < G_0(x_{n+1}) - G_0(y_{n-1})$. This completes the proof of (4).

5.8. PROPOSITION. Let C be a closed set, $U = R \setminus C$, $f_0, g_0 \in \mathcal{D}$ ($\mathcal{D}^+$ resp.), $\|f_0\| + \|g_0\| < \infty$, $\varphi \in \ell oc\, \mathcal{R}(U)$ ($\ell oc\, \mathcal{R}^+(U)$ resp.) and let $\eta \in (0, \infty)$. Then there are $f, g \in \mathcal{D}$ ($\mathcal{D}^+$ resp.) with $f = f_0$, $g = g_0$ on C, $fg = \varphi$ on U and

(9) $$\|f\|^p + \|g\|^p < \eta + 8^{p-1}(\|f_0\|^p + \|g_0\|^p) .$$

PROOF. Let $B = 8^{p-1}$. There is a $\delta \in (0, \infty)$ with $B((\|f_0\|_U + \delta)^p + (\|g_0\|_U + \delta)^p) < \frac{\eta}{2} + B(\|f_0\|_U^p + \|g_0\|_U^p)$. Choose an $\omega \in \Delta$ with $\omega > 0$ on U, $\omega = \omega' = 0$ on C and $\|\omega\| < \delta$. Let $F_0' = f_0 + \omega$, $G_0' = g_0 + \omega$. If $f_0, g_0 \in \mathcal{D}^+$, then $F_0'G_0' > 0$ on U. Applying 5.7 (5.7.1 resp.) to each component of U we get $F, G \in \Delta(U)$ ($\Delta^+(U)$ resp.) such that $F'G' = \varphi$, $|F - F_0| + |G - G_0| < \omega$ on U and $\|F'\|_U^p + \|G'\|_U^p < \frac{\eta}{2} + B(\|F_0'\|_U^p + \|G_0'\|_U^p)$. Since $\|F_0'\|_U < \|f_0\|_U + \delta$ and $\|G_0'\|_U < \|g_0\|_U + \delta$, we have

(10) $$\|F'\|_U^p + \|G'\|_U^p < \eta + B(\|f_0\|_U^p + \|g_0\|_U^p) .$$

Define $F = F_0$ and $G = G_0$ on C. Since $\omega = \omega' = 0$ on C, we have $F, G \in \Delta$ and $F' = F_0'$, $G' = G_0'$ on C. Let $f = F'$ and $g = G'$. Then $F' = f_0$, $G' = g_0$ on C and (9) follows at once from (10). If $f_0, g_0 \in \mathcal{D}^+$, then $f, g \in \mathcal{D}^+$ as well.

5.9. THEOREM. Let φ be a function on R. Suppose that for each nonempty closed set, C, there is an open interval I such that $C \cap I \neq \emptyset$ and $\varphi \in \mathcal{R}(C \cap I)$ ($\mathcal{R}^+(C \cap I)$ resp.). Then $\varphi \in \mathcal{R}$ ($\mathcal{R}^+$ resp.).

PROOF. Let $U = \{x :$ there is an open interval, I, with $x \in I$ and $\varphi \in \mathcal{R}(I)\}$, and let $C = R \setminus U$. Our objective is to show that $C = \emptyset$, for if so, then by 5.8 $\varphi \in \mathcal{R}$. So suppose $C \neq \emptyset$. Then there is an open interval, I, with $C \cap I \neq \emptyset$ and $\varphi \in \mathcal{R}(C \cap I)$. Let $\varepsilon \in (0, \infty)$. Then there are $f_0, g_0 \in \mathcal{D}$ such that $8^{p-1}(\|f_0\|^p + \|g_0\|^p) < \varepsilon$ and $f_0 g_0 = \varphi$ on $C \cap I$. By 5.8 there are $f, g \in \mathcal{D}$ such that $f = f_0$,

$g = g_0$ on C, $fg = \varphi$ on U, and $\|f\|^p + \|g\|^p < \epsilon$. Since $fg = \varphi$ on $(C \cap I) \cup U$, we have $\varphi \in \mathcal{R}(I)$ and hence $I \subset U$ contrary to $C \cap I \neq \emptyset$. Consequently $C = \emptyset$. The alternate assertion can be proved in the same way.

We finally come to the objective of this section which, with the help of 5.2, is an easy consequence of 5.9.

5.10. COROLLARY. Let φ be a Baire one function with $\varphi = 0$ a.e. on R. Then $\varphi \in \mathcal{R}$. If in addition $\varphi \geq 0$, then $\varphi \in \mathcal{R}^+$.

PROOF. Let C be a nonempty, closed set. There is an open interval, I, such that $C \cap I \neq \emptyset$ and φ is bounded on $S = C \cap I$. By 5.2 $\varphi\chi_S \in \mathcal{R}$. So by definition $\varphi \in \mathcal{R}(C \cap I)$. Thus 5.9 implies that $\varphi \in \mathcal{R}$. The proof of the additional assertion is similar.

BIBLIOGRAPHY

1. S. J. Agronsky, R. Biskner, A. M. Bruckner and J. Mařík, "Representations of functions by derivatives," Trans. Amer. Math. Soc. 263 (1981), 493-500.

2. R. J. Fleissner, "Multiplication and the fundamental theorem of calculus: A survey, Real Analysis Exchange, 2, No. 1 (1976), 7-34.

3. C. Kuratowski, Topologie, Vol. 1 (4th French edition), Warszawa, 1958.

4. J. Mařík and C. E. Weil, "Products of powers of non-negative derivatives," to appear in TAMS.

5. G. Petruska, M. Laczkovich, "Baire one functions, approximately continuous functions and derivatives," Acta Math. Acad. Sci. Hung., 25 (1-2), (1974), 189-212.

6. Z. Zahorski, "Sur la première dérivée," Trans. Amer. Math. Soc. 69 (1950), 1-54.

DEPARTMENT OF MATHEMATICS
UNIVERSITY OF CALIFORNIA-SANTA BARBARA
SANTA BARBARA, CALIFORNIA 93106

DEPARTMENT OF MATHEMATICS
MICHIGAN STATE UNIVERSITY
EAST LANSING, MICHIGAN 48824

Contemporary Mathematics
Volume 42, 1985

MONOTONE APPROXIMATION ON AN INTERVAL

Richard Darst and Robert Huotari

Let f be a bounded Lebesgue measurable function defined on an interval (say [0, 2]). For $1 < p < \infty$ there is a unique (a.e.) best L_p approximation f_p to f by monotone (non-decreasing) functions. We say that f has the Polya property if $f_\infty = \lim_{p\to\infty} f_p$ is well defined as a bounded measurable function: if $p_n \to \infty$, then $\lim_n f_{p_n}$ exists a.e. (on [0, 2]). If f is continuous, then f_p is continuous and $f_p \to f_\infty$ uniformly (on [0, 2]) [1]. If f is quasi-continuous, then f_p is quasi-continuous and $f_p \to f_\infty$ uniformly [1]. An example of a bounded measurable function which does not have the Polya property is given in [2]. Our purpose is to give an example of a bounded function f defined on [0, 2] which is continuous on [0, 1), continuous on [1, 2], approximately continuous at one and does not have the Polya property.

To begin, put $f = 5$ on [1, 2]. We will use a sequence $\{u_n, v_n\}$ of pairs of points, $0 = u_1 < v_1 < u_2 < \ldots \to 1$, to define f on [0, 1). To help explain, we record some properties that will be satisfied;
$(v_n - u_n) < 4^{-n}(1 - u_n)$, $(1 - u_n+1) < 4^{-n}(1 - u_n)$, $f(x) = 10$ on $[u_{2n-1}, v_{2n-1}]$, $f(x) = 0$ on $[u_{2n}, v_{2n}]$ and $f(x) \neq 5$ on less than 4^{-n} percent of $[v_n, u_{n+1}]$.

The following two facts will be used repeatedly. If $g \le h$, then $g_p \le h_p$, $1 < p < \infty$, and the map $g \to g_p$ is continuous in L_p, $1 < p < \infty$.

To start the definition of f on [0, 1), choose $v_1 < 4^{-1}$. Put $f = 10$ on $[0, v_1]$. Proceeding in steps, we will define f temporarily on $(v_1, 1)$, modify the temporary definition ϕ of f and then define f on one more piece, $[v_1, u_2]$, of [0, 1). Put $\phi = 5$ on $(v_1, 1)$ and $\phi = f$ elsewhere. Begin to increase p from 1. As p increases, $\phi_p \equiv v_p$ increases from 5 to 7.5. Choose $p_1 > 1$ with $\phi_{p_1} > 7$. Put $\phi^t = \phi - 5I_{(t,1)}$, $v_1 < t < 1$, where $I_E(x) = 1$ if $x \in E$ and $I_E(x) = 0$ otherwise. Choose u_2 with $(1 - u_2) < 4^{-1}(1 - u_1)$ and $\phi^{u_2}_{p_1} > 7$. Modify ϕ^{u_2} on less than 4^{-1} percent of $[v_1, u_2]$

1980 Mathematics Subject Classification. 41A30.

so that it decreases continuously from 10 to 0 and retains the property: $\phi_{p_1}^{u_2} > 7$. Put $f = \phi^{u_2}$ on $[v_1, u_2]$. Since $\phi^{u_2} = 0$ on $(u_2, 1)$, $f_{p_1} > 7$ if $f \geq 0$ on $(u_2, 1)$.

To continue the definition of f, choose v_2 with $(v_2 - u_2) < 4^{-2}(1 - u_2)$. Put $f = 0$ on u_2, v_2 . Put $\phi = 5$ on $(v_2, 1)$ and $\phi = f$ elsewhere. Begin to increase p from p_1. As p increases, ϕ_p decreases to 5 on 0, 2 . Choose $p_2 > 2p_1$ with $\phi_{p_2} < 6$. Put $\phi^t = \phi + 5I_{(t,1)}$, $v_2 < t < 1$. Choose u_3 with $(1 - u_3) < 4^{-2}(1 - u_2)$ and $\phi_{p_2}^{u_3} < 6$. Modify ϕ^{u_3} on less than 4^{-2} percent of $[v_2, u_3]$ so that it increases continuously from 0 to 10 and retains the property: $\phi_{p_2}^{u_3} < 6$. Put $f = \phi^{u_3}$ on $[v_2, u_3]$. Since $\phi^{u_3} = 10$ on $(u_3, 1)$, $f_{p_2} < 6$ if $f \leq 10$ on $(u_3, 1)$.

One facet of the construction remains to be displayed, so we begin one more step. Choose v_3 with $(v_3 - u_3) < 4^{-3}(1 - u_3)$. Put $f = 10$ on $[u_3, v_3]$. Put $\phi = 5$ on $(v_3, 1)$ and $\phi = f$ elsewhere. Begin to increase p from p_2. As p increases, ϕ increases to 7.5 on $[u_3, 2] \supset [1, 2]$. Continuing our procedure produces a function f with the promised properties: $f_{p_{2n-1}} > 7$ on $[1, 2]$ and $f_{p_{2n}} < 6$ on $[1, 2]$.

BIBLIOGRAPHY

1. R. B. Darst and Salem Sahab, "Approximation of continuous and quasi-continuous functions by monotone functions", J. Approx. Theory 38(1983), 9-27.

2. R. B. Darst, D. A. Legg and D. W. Townsend, "The Polya algorithm in L_∞ approximation, J. Approx. Theory 38(1983), 209-220.

DEPARTMENT OF MATHEMATICS
COLORADO STATE UNIVERSITY
FORT COLLINS, CO 80523

Contemporary Mathematics
Volume 42, 1985

TWO REMARKS ON THE MEASURE OF PRODUCT SETS

Roy O. Davies

ABSTRACT. It is shown that if every union of less than c null sets is a null set, then a measurable set of full measure in R^2 contains the product of two sets of full outer measure in R. A product set characterization of Sierpiński's property (C) is given.

1. MEASURABLE PLANE SETS OF FULL MEASURE CONTAIN LARGE PRODUCT SETS. Gillis [8] constructed a measurable set of full measure in the plane containing no measurable product set $X \times Y$ with the linear measures $\Lambda(X)$, $\Lambda(Y)$ both positive; Eggleston [5], and independently Erdös and Oxtoby [6], gave the following simple example: remove from the plane the set D of all points on lines $y = x+r$ with r rational. (This is simpler also than the example published later by Darst and Goffman [4].) Some positive results, when weaker requirements are imposed on X and Y, are given in [5] and by Brodskii [3] and Taylor [11].

Gillis asserted that his set contains no product $X \times Y$ with the outer measures $\Lambda^*(X)$, $\Lambda^*(Y)$ both positive. However he evidently noticed a gap in the argument, because a footnote states "*This last step actually requires proof. . . .the details . . . can easily be supplied by the reader. . . ." Eggleston asserted that his set $R^2\backslash D$ also contains no product $X \times Y$ with $\Lambda^*(X) > 0$ and $\Lambda^*(Y) > 0$. However, this is false, because with the help of the axiom of choice it is easy to construct a set X with inner measure 0 and full outer measure, such that whenever $x \in X$ and r is rational $x+r \in X$; now $X \times (R\backslash X)$ is contained in $R^2\backslash D$. What is true is that if $R^2\backslash D$ contains $X \times Y$ and $\Lambda^*(X) > 0$, $\Lambda^*(Y) > 0$, then both X and Y must have inner measure zero; this follows immediately from the well-known refinement of Steinhaus's theorem which asserts that the set of differences of two sets contains an

---

1980 Mathematics Subject Classification. 28A35.

interval if one of the sets has positive inner measure and the other has positive outer measure.

The question therefore arises: is a set of the type proposed by Gillis possible at all? It seems not to have been noticed before that it is easy to give a negative answer, if we assume that every union of less than $\underline{c}$ null-sets is a null-set, which follows from the continuum hypothesis or Martin's Axiom.

THEOREM 1. If every union of less than c null-sets is a null-set, then every measurable set E of full measure in the plane contains a product-set $X \times Y$ with X and Y of full outer measure in $\mathbb{R}$.

On the other hand, with the help of the axiom of choice and transfinite induction it is very easy to construct a non-measurable plane set of full outer measure with at most one point on each horizontal and vertical line; such a set contains no product sets other than singletons.

PROOF OF THEOREM 1. List all closed subsets of $\mathbb{R}$ of positive measure as K_α $(0 \leq \alpha < \omega_c)$, and choose $x_\alpha, y_\alpha \in \mathbb{R}$ by transfinite induction as follows. Suppose that x_β, y_β have been constructed for all $\beta < \alpha$, and that $\{x_\beta : \beta < \alpha\} \times \{y_\beta : \beta < \alpha\} \subseteq E$ and for all $\beta < \alpha$ the (linear) sets $\{y : (x_\beta, y) \in E\}$ and $\{x : (x, y_\beta) \in E\}$ are both measurable and of full measure, and $x_\beta, y_\beta \in K_\beta$. Then for x_α we choose any element $x \in K_\alpha$ with the two properties: (a) $(x, y_\beta) \in E$ for all $\beta < \alpha$, (b) $\{y : (x,y) \in E\}$ is measurable and of full measure. For y_α we then choose any element $y \in K_\alpha$ with the two properties: (a) $(x_\beta, y) \in E$ for all $\beta < \alpha$, (b) $\{x : (x,y) \in E\}$ is measurable and of full measure. Let $X = \{x_\alpha : 0 \leq \alpha < \omega_c\}$, $Y = \{y_\alpha : 0 \leq \alpha < \omega_c\}$.

2. PROPERTY (C) FOR $E \iff \Lambda(E \times F) = 0$ FOR EVERY COMPACT NULL-SET F.

A linear set E has property (C) (according to the terminology of Sierpiński [9], [10]) if for every sequence of positive numbers $\delta_1, \delta_2, \ldots$ it can be covered by a sequence of intervals with respective lengths δ_n. This idea has aroused intermittent interest since its inception by Borel [2] in 1919; with the help of (for example) the continuum hypothesis it is possible to construct an uncountable set possessing property (C). Of particular interest is the recent discovery [7] that E has property (C) if and only if every dense G_δ set contains a translate of E. Here we present a product-set characterization.

One half of this result is essentially contained in [11], but since it is not stated explicitly there, and relies also on a result of Besicovitch [1], it is worth summarizing Taylor's argument.

PROPOSITION. If E does not have property (C) then there is a compact null-set F with $\Lambda(E \times F) > 0$.

SKETCH OF PROOF. There exist positive numbers $\delta_1, \delta_2, \ldots$ such that E cannot be covered by a sequence of intervals with respective lengths δ_n. We may suppose that $n\delta_n \downarrow 0$. Define $\phi(x)$ as follows: $\phi(\delta_n) = 1/n$ for $n = 1, 2, \ldots$, and ϕ is linear between these points; $\phi(0) = 0$; and $\phi(x) = 1 + (x - \delta_1)$ for $x > \delta_1$. This is continuous and strictly increasing on $[0, \infty)$, with $\phi(0) = 0$, and moreover the 'conjugate function' $\psi(x) = x/\phi(x)$ has the same properties (if we put $\psi(0) = 0$). The associated Hausdorff measures Λ^ϕ, Λ^ψ are well defined, and $\Lambda^\phi(E) \geq 1$. By a Cantor-style construction we can find a compact linear set F such that $1 \leq \Lambda^\psi(F) < \infty$. This set has linear measure zero because $\Lambda^\psi(F) < \infty$, and $\Lambda(E \times F) \geq 1$ because $\Lambda^\psi(F) \geq 1$.

The following result completes the proof of our characterization.

THEOREM 2. If E has property (C) then $\Lambda(E \times F) = 0$ for every compact null-set F.

PROOF. It is sufficient to show that given any $\epsilon > 0$, there exists a covering of $E \times F$ by a countable collection of sets, $\mathcal{C}$, with sum of diameters less than ϵ.

We can choose positive integers $n_1 < n_2 < \ldots$, and positive integers $m_1, m_2, \ldots$ such that F can be covered by a collection $\mathcal{J}_i$ of m_i intervals of length $1/2^{n_i}$ and

$$(m_i/2^{n_i})\sqrt{2} < \epsilon/2^i .$$

Since E has property (C), it can be covered by a sequence of intervals J_i with respective lengths $1/2^{n_i}$. For each interval $K \in \mathcal{J}_i$ the square $J_i \times K$ has diameter $(1/2^{n_i})\sqrt{2}$, and there are m_i of these squares, so the sum of their diameters if $(m_i/2^{n_i})\sqrt{2}$. We can now take as $\mathcal{C}$ the collection of all such squares for $i = 1, 2, \ldots$.

I am grateful to H. G. Eggleston, R. A. Johnson, J. C. Oxtoby, and S. J. Taylor for helpful comments.

BIBLIOGRAPHY

[1] A. S. Besicovitch, "Concentrated and rarified sets", Acta Math. 62 (1934), 289-300.

[2] E. Borel, "Sur la classification des ensembles de mesure nulle", Bull. Soc. Math. France 47 (1919), 97-125.

[3] M. L. Brodskii, "On some properties of sets of positive measure", Uspekhi Mat. Nauk 4, no. 3(31) (1949), 136-139. (Russian)

[4] R. B. Darst and C. Goffman, "A Borel set which contains no rectangles", Amer. Math. Monthly 77 (1970), 728-729.

[5] H. G. Eggleston, "Two measure properties of Cartesian product sets", Quart. J. Math. (Oxford) (2) 5 (1954), 108-115.

[6] P. Erdős and J. C. Oxtoby, "Partitions of the plane into sets having positive measure in every non-null measurable product set". Trans. Amer. Math. Soc. 79 (1955), 91-102.

[7] F. Galvin, J. Mycielski, and R. Solovay, "Strong measure zero sets", Notices Amer. Math. Soc. 26 (1979), A-280.

[8] J. Gillis, "Some combinatorial properties of measurable sets", Quart. J. Math. (Oxford) 7 (1936), 191-198.

[9] W. Sierpiński, "Sur un ensemble non dénombrable, dont tout image continue est de mesure nulle", Fund. Math. 11 (1928), 302-304; Oeuvres choisies, tome II (Warsaw 1975), 702-704.

[10] W. Sierpiński, "Sur le rapport de la propriété (C) à la théorie générale des ensembles", Fund. Math. 29 (1937), 91-96; Oeuvres choisies, tome III (Warsaw 1976), 351-355.

[11] S. J. Taylor, "On Cartesian product sets", J. London Math. Soc. 27 (1952), 295-304.

DEPARTMENT OF MATHEMATICS
THE UNIVERSITY
LEICESTER
LE1 7RH
ENGLAND

Contemporary Mathematics
Volume 42, 1985

MONOTONICITY, SYMMETRY, AND SMOOTHNESS

Michael J. Evans and Lee Larson

ABSTRACT. It is well known that a real valued continuous function which possesses a non-negative right upper Dini derivate at each point of the real line must be nondecreasing. Here it is shown that this result remains valid for measurable functions if the continuity assumption is replaced by smoothness and that it does not remain valid if this assumption is replaced by symmetry.

A real valued function f defined on the real line R is said to be _symmetric_ [_smooth_] if at each $x \in R$, $\Delta^2 f(x,h) = o(1)$ [$o(h)$] as $h \to 0$, where $\Delta^2 f(x,h) = f(x+h) + f(x-h) - 2f(x)$. Various continuity and differentiability properties of measurable symmetric and smooth functions have been studied extensively in [1,2,3,5]. In the present paper we wish to prove that the familiar result which states that a continuous function, f, possessing a non-negative right upper Dini derivate, $D^+f(x)$, at each $x \in R$ must be nondecreasing remains valid if continuity is replaced by smoothness, but is not valid if continuity is replaced by symmetry.

We begin by stating an elementary lemma, which is generally well-known. (See Theorem 7.2 on page 204 of [4], for example.)

LEMMA. _If a real valued function_ f _satisfies both conditions_

i) $D^+f(x) \geq 0$

and

ii) $\limsup_{t \to x^-} f(t) \leq f(x)$

at each point x _of an interval_ I, _then_ f _is nondecreasing on_ I.

THEOREM. _If_ $f: R \to R$ _is measurable and smooth with_ $D^+f(x) \geq 0$ _everywhere, then_ f _is nondecreasing_.

1980 Mathematics Subject Classification. 26A48, 26A15.

Proof. There is no loss of generality in assuming that $D^+f(x) > 0$ everywhere since we could replace $f(x)$ with $g(x) = f(x) + \varepsilon x$ for arbitrary positive ε.

We shall show that the set

$$E = \{x : \limsup_{t \to x^-} f(t) > f(x)\}$$

is empty. Suppose that E is nonempty. In [1] it was shown that the set of points at which a measurable smooth function is discontinuous is clairseme, or scattered; i.e., every nonempty subset of the set of points of discontinuity has an isolated point. Consequently, E has an isolated point, x_0. Choose $\delta > 0$ such that $(x_0-\delta, x_0) \cap E = \phi$. According to the lemma, f is nondecreasing on $(x_0-\delta, x_0)$.

Since $x_0 \in E$ and f is nondecreasing on $(x_0-\delta, x_0)$, there is an $\alpha > 0$ and a δ' with $0 < \delta' \leq \delta$ such that $f(t) > f(x_0) + \alpha$ for $t \in (x_0-\delta', x_0)$. Since f is smooth at x_0, there is a δ^* with $0 < \delta^* \leq \delta'$ such that $|\Delta^2 f(x_0,h)| < \alpha$ for $0 < h < \delta^*$. Hence, for $0 < h < \delta^*$ we have

$$\begin{aligned} f(x_0 + h) &= \Delta^2 f(x_0,h) + (f(x_0) - f(x_0-h)) + f(x_0) \\ &< \alpha - \alpha + f(x_0) \\ &= f(x_0) , \end{aligned}$$

contradicting the assumption that $D^+f(x_0) > 0$, and completing the proof.

EXAMPLE. There is a measurable function $f: (0,1) \to R$ such that

i) f is symmetric,

ii) $D^+f(x) \geq 0$ everywhere,

and

iii) f is not monotone.

Proof. We shall construct a sequence of functions f_n, $n = 0,1,\ldots,$ having the following properties:

a_n) For each $k = 1,2,\ldots,$ $f_n(1/2^{2k}) = 1$ and $f_n(1/2^{2k+1}) = 0$.

b_n) f_n is continuous except at m_n points $z_1,z_2,\ldots,z_{m_n}$, where $m_n = \sum_{k=1}^{n} (2^k-1)$.

c_n) f_n is symmetric at each of $z_1,z_2,\ldots,z_{m_n}$.

d_n) f_n is decreasing and linear on each of a sequence of pairwise disjoint closed intervals, I_k^n, $k = 1,2,\ldots$; f_n is nondecreasing and piecewise linear on a sequence of pairwise disjoint closed intervals, J_k^n, $k = 1,2,\ldots$; and if $\mathcal{I}^n = \bigcup_{k-1}^{\infty} I_k^n$ and $\mathcal{J}^n = \bigcup_{k=1}^{\infty} J_k^n$, then $(0,1) = \mathcal{I}^n \cup \mathcal{J}^n \cup \{z_1,z_2,\ldots,z_{m_n}\}$.

e_n) For each $k = 1,2,\ldots$, and each $x \in I_k^n$, there is a number h such that $0 < h < 3|I_k^n|$, $x + h \in \mathcal{J}^n$ and $f_n(x+h) \geq f_n(x)$.

f_n) For each z_i, $i = 1,2,\ldots,m_n$, there is a sequence of positive numbers, t_j^i, $j = 1,2,\ldots$, such that $z_i + t_j^i \in \mathcal{J}^n$,

$f_n(z_i + t_j^i) = f_n(z_i) = f_{n-1}(z_i)$ for each sufficiently large natural number j and $\lim_{j\to\infty} t_j^i = 0$.

g_n) $\mathcal{J}^n \supseteq \mathcal{J}^{n-1}$ and $\mathcal{I}^n \subseteq \mathcal{I}^{n-1}$.

h_n) For each $x \in \mathcal{J}^n$, $f_n(x) \geq f_{n-1}(x)$.

i_n) For each $x \in (0,1)$, $|f_n(x)-f_{n-1}(x)| < 1/2^n$

and

j) $\lim_{n\to\infty} (\max \{|I_k^n|: k = 1,2,\ldots\}) = 0$, where $|I|$ denotes the length of an interval I.

For the moment, suppose that such a function sequence, $\{f_n\}$, exists. From i_n) we have $\{f_n\}$ converging uniformly to a function f on $(0,1)$. From b_n) and c_n) we see that each f_n is symmetric on $(0,1)$ and it is easily shown that the uniform limit of a sequence of symmetric functions is symmetric. From conditions d_n), e_n), f_n), g_n), h_n), and j) it is easily shown that $D^+f(x) \geq 0$ for each $x \in (0,1)$. However, condition a_n) shows that f is not monotone.

We now turn to the construction of this function sequence $\{f_n\}$. We begin by defining a function α on $(0,1]$, which will serve as the "building block" of the construction. For $k = 0,1,\ldots$, let $\alpha(1/2^{2k}) = 1$ and $\alpha(1/2^{2k+1}) = 0$, and then extend α linearly on $(0,1]$. We let ℓ, r, β, γ, and g denote functions defined as follows:

$$\ell(x) = \alpha(1-x), \qquad 0 < x < 1,$$

$$r(x) = 1 - \alpha(x), \qquad 0 < x < 1,$$

$$\beta(x) = \begin{cases} r(x), & 0 < x \leq 1/2, \\ 1 - \ell(x), & 1/2 < x < 1, \end{cases}$$

$$\gamma(x) = 1 - \beta(x), \qquad 0 < x < 1$$

and

$$g(x) = \begin{cases} 2x, & 0 \leq x \leq 1/2 \\ 1, & 1/2 < x \end{cases}$$

We start by letting $f_0 = \alpha$ on $(0,1)$. Then f_0 is continuous on $(0,1)$, and satisfies conditions a_0), b_0), and c_0), the latter two vacuously. For each $k = 1,2,\ldots$, let $I_k^0 = [1/2^{2k}, 1/2^{2k-1}]$ and $J_k^0 = [1/2^{2k-1}, 1/2^{2k-2}]$. It is easy to see that d_0) and e_0) are satisfied; f_0), g_0), h_0), and i_0) are vacuous.

To obtain the function f_1, we shall alter f_0 on the longest of the intervals I_k^0, $k = 1,2,\ldots$, namely, $I_1^0 = [1/4, 1/2]$, as well as on a relatively short interval at the left end of $J_1^0 = [1/2, 1]$. We set

$$f_1(x) = \begin{cases} \ell(8(x-1/4))/2 + 1/2, & x \in (1/4,3/8), \\ r(8(x-3/8))/2, & x \in (3/8,1/2), \\ g(4(x-1/2))/2, & x \in (1/2,3/4), \\ f_0(x), & \text{otherwise.} \end{cases}$$

It is easy to see that f is continuous everywhere except at $z_1 = 3/8$. If $0 < t < 1/3$, then

$$\begin{aligned} \Delta^2 f_1(3/8,t) &= r(8t)/2 + \ell(8(1/8-t))/2 + 1/2 - 1 \\ &= (1-\alpha(8t))/2 + \alpha(8t)/2 - 1/2 \\ &= 0 . \end{aligned}$$

Hence f_1 is symmetric at z_1, and a_1), b_1), and c_1) are satisfied.

Let $\{I_k^1\}_{k=1}^\infty$ denote the sequence of maximal closed intervals on which f_1 is decreasing; i.e., f_1 is decreasing on I_k^1, but on no larger interval containing I_k^1. Note that for each k, $|I_k^1| = 1/2^{i_k}$ for some natural number $i_k \geq 4$. We let $\{J_k^1\}_{k=1}^\infty$ denote the sequence of maximal closed intervals on each of which f_1 is nondecreasing. Clearly, $(0,1) = \mathcal{I}^1 \cup \mathcal{J}^1 \cup \{z_1\}$, satisfying d_1).

Let $I_{k_0}^1$ be that particular interval of the sequence $\{I_k^1\}$ whose right endpoint is 1/2; i.e., $I_{k_0}^1 = [7/16,1/2]$. If $x \in I_{k_0}^1$ and we set $h = 5/8 - x$, then $0 < h \leq 3|I_{k_0}^1| = 3/16$, $x + h = 5/8 \in \mathcal{J}^1$, and $f_1(x+h) = 1/2 \geq f_1(x)$. (This is why we modified f_0 on the left end of J_1^0.) For x in any other I_k^1 it is obvious that the requirement of e_1) is met.

To satisfy condition f_1), define the sequence $\{t_j^1\}_{j=1}^\infty$ by $t_j^1 = 1/2^{2j-1}$. Also, conditions g_1), h_1), and i_1) are clearly satisfied.

Before moving to the inductive step, we find it instructive to treat the n=2 case explicitly as well. We let $I^2 = [a,b]$ denote the longest of the intervals in $\{I_k^1\}_{k=1}^\infty$. If there is a tie for this distinction, we let I^2 be the right-most such interval. We let J^2 denote that member of $\{J_k^1\}_{k=1}^\infty$ whose left endpoint is b. (For the reader interested in sketching the first few functions we note that $I^2 = [1/4,5/16]$ and $J^2 = [5/16,11/32]$.) We partition I^2 into four intervals of equal length $L = (b-a)/4$ at the points $a = x_0 < x_1 < x_2 < x_3 < x_4 = b$. We let $H = (f_1(a)-f_1(b))/4 = f_1(x_k)-f_1(x_{k-1})$, $k = 1,2,3,4$ and define f_2 by

$$f_2(x) = \begin{cases} H \cdot \ell((x - x_0)/L) + f_1(x_1), & x \in (x_0,x_1), \\ H \cdot \beta((x - x_1)/L) + f_1(x_2), & x \in (x_1,x_2), \\ H \cdot \gamma((x - x_2)/L) + f_1(x_3), & x \in (x_2,x_3), \\ H \cdot r((x - x_3)/L) + f_1(x_4), & x \in (x_3,x_4), \\ \max\{H\cdot g((x-x_4)/2L)+f_1(x_4),\ f_1(x)\}, & x \in J^2, \\ f_1(x), & \text{otherwise}. \end{cases}$$

Setting $z_2 = x_1$, $z_3 = x_2$, and $z_4 = x_3$, a_2) and b_2) are clearly satisfied. It is routine to verify that f_2 is symmetric at z_1, z_2, z_3, and z_4. As a sample of the calculations, we verify that f_2 is symmetric at z_2 as follows: For $0 < t < L/2$.

$$\begin{aligned} \Delta^2 f_2(z_2,t) &= H \cdot \beta(t/L) + f_1(x_2) + H \cdot \ell(1-t/L) + f_1(z_2) - 2f_1(z_2) \\ &= H \cdot (1-\alpha(t/L)) + f_1(x_2) + H \cdot \alpha(t/L) - f_1(z_2) \\ &= H + f_1(x_2) - f_1(z_2) \\ &= f_1(z_2) - f_1(z_2) \\ &= 0. \end{aligned}$$

Consequently, f_2 is symmetric at z_2. Similarly, symmetry holds at z_3 and z_4. Since the distance between z_1 and $I^2 \cup J^2$ is positive, there is an open interval containing z_1 on which $f_2 = f_1$, yielding the symmetry of f_2 at z_1. Thus c_2) holds.

We let $\{I_k^2\}_{k=1}^{\infty}$ denote the sequence of maximal closed intervals on which f_2 is decreasing and $\{J_k^2\}_{k=1}^{\infty}$ denote the sequence of maximal closed intervals on which f_2 is nondecreasing. Then $(0,1) = \mathcal{I}^2 \cup \mathcal{J}^2 \cup \{z_1,z_2,z_3,z_4\}$ to satisfy condition d_2). Condition e_2) follows in the same manner as e_1).

For $i=2$ and 4 we let $t_j^i = 1/2^{2j-1}$, $j = 1,2,\ldots$, and $t_j^3 = 1/2^{2j}$, $j = 1,2,\ldots$. With these sequences, condition f_2) follows.

Conditions g_2), h_2), and i_2) are obviously satisfied.

Suppose now that $f_0, f_1, \ldots, f_n$ have been defined in this manner with f_k satisfying conditions a_k) through i_k) for each $k = 0,1,\ldots,n$. We now define f_{n+1} in much the same fashion that f_2 was defined. We let $I^{n+1} = [a,b]$ denote the longest of the intervals in $\{I_k^n\}_{k=1}^{\infty}$. Again, if there is a tie for this distinction, we let I^{n+1} be the right-most such interval. We let J^{n+1} denote that member of $\{J_k^n\}_{k=1}^{\infty}$ whose left endpoint is b.

Let $L = (b-a)/2^{n+1}$ and $x_k = a+kL$, $k = 0,1,\ldots,2^{n+1}$. Let $H = (f_n(a)-f_n(b))/2^{n+1}$. Then $H \le 1/2^{n+1}$ and $H = f_n(x_k)-f_n)x_{k-1})$, $k = 1,2,\ldots,2^{n+1}$. Let

$$f_{n+1}(x) = \begin{cases} H\cdot\ell(x-x_0)/L + f_n(x_1), & x \in (a,a+L), \\ H\cdot\beta((x-x_k)/L) + f_n(x_{k+1}), & x \in (x_k,x_{k+1}),\ k=2i-1,\ i=1,2,\ldots,2^n-1, \\ H\cdot\gamma((x-x_k)/L) + f_n(x_{k+1}), & x \in (x_k,x_{k+1}),\ k=2i,\ i=1,2,\ldots,2^n-1, \\ H\cdot r((x-b+L)/L) + f_n(b), & x \in (b-L,b), \\ \max\{H\cdot g((x-b)/2L) + f_n(b),\ f_n(x)\}, & x \in J^{n+1}, \\ f_n(x), & \text{otherwise.} \end{cases}$$

For $i = m_n+1, m_n+2,\ldots, m_n+2^{n+1}-1 = m_{n+1}$ let $z_i = x_{i-m_n}$. For $i = 1,2,\ldots,m_n$ the distance from z_i to $I^{n+1} \cup J^{n+1}$ is positive and so $f_{n+1} = f_n$ on an open interval about z_i, yielding the symmetry of f_{n+1} at z_i. For $i = m_n+1,\ldots,m_n+2^{n+1}-1$ the symmetry of f_{n+1} at z_i follows from calculations analogous to those presented in the $n=1$ and $n=2$ cases. Thus a_{n+1}), b_{n+1}), and c_{n+1}) are satisfied.

As before we let $\{I_k^{n+1}\}_{k=1}^{\infty}$ denote the sequence of maximal closed intervals on which f_{n+1} is decreasing and $\{J_k^{n+1}\}_{k=1}^{\infty}$ denote the sequence of maximal closed intervals on which f_{n+1} is nondecreasing. Then $(0,1) = \mathcal{J}^{n+1} \cup \mathcal{I}^{n+1} \cup \{z_1,z_2,\ldots,z_{m_{n+1}}\}$ and condition d_{n+1}) holds.

Condition e_{n+1}) is verified in the same manner as was e_1).

For $i = m_n+2p$, $p = 1,2,\ldots,2^n-1$, let $t_j^i = 1/2^{2j}$, $j = 1,2,\ldots$ and for all other i such that $m_n+1 \le i \le m_{n+1}$, let $t_j^i = 1/2^{2j-1}$, $j = 1,2,\ldots$. With these sequences, the condition f_{n+1}) is easily checked.

Conditions g_{n+1}), h_{n+1}), and i_{n+1}) again follow routinely.

Finally, to see that j) holds, simply note that the sequence $\{\max\{|I_k^n| : k = 1,2,\ldots\}\}_{n=0}^{\infty}$ is nonincreasing and that for each n there is an $m > n$ such that

$$\max\{|I_k^m| : k = 1,2,\ldots\} \le \max\{|I_k^n| : k = 1,2,\ldots\}/2.$$

BIBLIOGRAPHY

1. M. J. Evans and L. Larson, "The continuity of symmetric and smooth functions", Acta Math. Acad. Sci. Hungaricae (to appear).

2. C. J. Neugebauer, "Smoothness and differentiability in L_p", Studia Math. 25 (1964), 81-91.

3. ________, "Symmetric, continuous and smooth functions", Duke Math. J. 31 (1964), 23-32.

4. S. Saks, Theory of the Integral, Monografie Matematyczne 7, Warsawa-Lwow, 1937.

5. A. Zygmund, Trigonometric Series, vol. I-II, Cambridge, 1959.

DEPARTMENT OF MATHEMATICS
NORTH CAROLINA STATE UNIVERSITY
RALEIGH, NORTH CAROLINA 27650

DEPARTMENT OF MATHEMATICS
UNIVERSITY OF LOUISVILLE
LOUISVILLE, KENTUCKY 40292

Contemporary Mathematics
Volume 42, 1985

THE STRUCTURE OF CONTINUOUS FUNCTIONS WHICH SATISFY LUSIN'S CONDITION (N)

James Foran

ABSTRACT. Examples are given which illuminate the nature of Lusin's condition (N). Two characterizations of functions satisfying (N) are given. One is global in nature while the other involves behavior at points. An exceptional set of measure zero is allowed in each characterization.

A function F is said to satisfy Lusin's condition (N) if the image under F of each set of measure 0 is of measure 0. Lusin's condition (N) plays an important role in the theory of the integral. As is well known, for a continuous function F of bounded variation, Lusin's condition (N) is necessary and sufficient for F to be absolutely continuous; i.e., for F to be a Lebesgue integral. The condition also plays an analogous role for the Denjoy integral. It is, however, not known whether every integral which generalizes the Lebesgue integral must satisfy (N). Much of the early work on (N) can be found in [4] and in the references given there.

In a previous paper [2], it was shown that integrals must satisfy a condition slightly weaker than (N) which guarantees agreement with the Lebesgue integral. Namely, they must satisfy (M):

(M): A function F is said to satisfy (M) provided that whenever E is a set on which F is of bounded variation, F is absolutely continuous on E.

In [3] it was shown that (M) is equivalent to the condition that if E is a set on which F is monotone, F is also absolutely continuous on E. It was also shown that there are F which satisfy (M) but do not satisfy (N). An example was given of an $F \in (M)$ which did not satisfy Banach's condition T_2.

Mathematics Subject Classification. 26A15, 26A46.

T_2: A function F is said to satisfy T_2 if, for almost every y, $F^{-1}(y)$ is at most countable.

Continuous F which satisfy (N) also satisfy T_2 (see [4], p. 284]. The example below shows that there are continuous functions which satisfy both (M) and T_2 but do not satisfy (N).

EXAMPLE 1. A continuous function $F(x)$ defined on $[0,1]$ and satisfying (M) and T_2 but not (N).

CONSTRUCTION. Each point x in the Cantor ternary set C can be written uniquely as $\Sigma a_i/9^i$ where $a_i = 0,2,6$ or 8. For such x let $F(x) = \Sigma b_i/4^i$ where $b_i = 2$ if $a_i = 0$, $b_i = 1$ if $a_i = 2$, $b_i = 4$ if $a_i = 6$, and $b_i = 3$ if $a_i = 8$. Then $F(x)$ is continuous on C and can be extended linearly to a function F which is continuous on $[0,1]$. Since the image under F of C is $[0,1]$, F does not satisfy (N). Except for the countable collection of points $x \in C$ where $F(x) = \Sigma b_i/4^i$ has two representations (one with $b_i = 3$ for $i > N$, one with $b_i = 0$ for $i > N$) $F(x)$ is one to one on C. Because there are countably many line segments on the graph of F, it follows that for each $y \in [0,1]$ $F^{-1}(y)$ is at most countable. Thus F satisfies T_2. Since F satisfies (M) on the complement of C, it remains to show that F satisfies (M) on C. However, if $E \subset C$ is a set on which F is monotone, E can contain points in at most two of the intervals $[0,1/9]$, $[2/9,3/9]$, $[6/9,7/9]$ and $[8/9,1]$. Thus $|F(E)| \leq 1/2$. Similarly, in each of these two intervals, E can contain points in at most two of the four subintervals determining C. Continuing in this fashion, it is clear that $|F(E)| \leq 1/2^n$ and thus $|F(E)| = 0$ and F satisfies (M).

Banach's condition T_1 (a function satisfies T_1 if, for almost every y, $F^{-1}(y)$ is finite) along with (M) does imply (N). This follows easily from the characterization of T_1 given in [4] p. 278. Namely, if a continuous function F satisfies T_1, then except on a set whose image has measure 0, F has either a finite or an infinite derivative. Then, on the complement of this set whose image has measure 0, F is of generalized bounded variation. This along with condition (M) implies generalized absolute continuity on this complement which in turn implies condition (N) (confer [4] p. 225). In fact, if on the complement of a set whose image is of measure 0 a function satisfies at each x either $\overline{F}_{ap}(x) < \infty$ or $\underline{F}_{ap}(x) > -\infty$, then F must be of generalized bounded variation on the complement and this together with (M) implies (N). Such an F is approximately derivable at almost every point of the complement. As the example below shows, this does not characterize (N).

EXAMPLE 2. A continuous function F defined on $[0,1]$ and satisfying (N) such that there is a set E with $|E|>0$ with F one to one on E and nowhere approximately derivable on E. The construction for Example 2 is similar to that for Example 1 but differs in that it utilizes a perfect set of positive measure which is mapped onto itself in the same oscillatory fashion.

CONSTRUCTION. For each natural number n, let $1/2>\epsilon_n>0$ be determined so that $\Pi(1-\epsilon_n)>0$. Subdivide the interval $[0,1]$ into four equal length intervals. Let I_1, I_2, I_3, I_4 be the closed intervals of length $4^{-1}(1-\epsilon_1)$ centered on each of the four equal length intervals. Let $E_1 = \cup I_i$ and let

$$H_1 = (I_1 \times I_2) \cup (I_2 \times I_1) \cup (I_3 \times I_4) \cup (I_4 \times I_3)$$

in general, if $I_{i_1,i_2,\ldots,i_n}$ $i_j = 1,2,3,4$ E_n and H_n have been defined, subdivide each $I_{i_1,i_2,\ldots,i_n}$ into four equal parts and let $I_{i_1,i_2,\ldots,i_n,j}$ be the four intervals centered on each quarter of $I_{i_1,i_2,\ldots i_n}$ and having length $4^{-n-1}(1-\epsilon_1)\cdots(1-\epsilon_n)(1-\epsilon_{n+1})$. Let $E_{n+1} = \cup I_{i_1,i_2,\ldots i_n,j}$ and let $H_{n+1} = H_n \cap \cup (I_{i_1,i_2,\ldots i_n,i}) \times (I_{j_1,j_2,\ldots j_n,j})$ where the union of the cross products extends over all indices such that $i\cdot j = 2$ or 12. Let $E = \cap E_n$ and define $F(x)$ on E so that $(x,F(x)) \in \cap H_n$. Since the points (x,y) of H_n above each $x \in E_n$ is an interval of length less than 4^{-n}, $F(x)$ is uniquely defined on E. Since the graph of F on E is $\cap H_n$ which is compact, $F(x)$ is continuous on E. Define $F(x)$ linearly on the complement of E. By noting that for any interval J $|F(E\cap J)| = |E\cap J|$, it follows that F satisfies (N) on E. F is also one to one on E. Let $x \in I_{i_1,i_2,\ldots i_n,j}$ and $t \in I_{i_1,i_2,\ldots i_n,k}$. Then

$$\text{if } j\cdot k = 2 \text{ or } 12, \quad \frac{f(t)-f(x)}{t-x} < 0$$

$$\text{if } j\cdot k = 3 \text{ or } 6, \quad \frac{f(t)-f(x)}{t-x} > 1/4 .$$

Since at each point $x \in E$ the intervals $I_{i_1,i_2,\ldots i_n,k}$ have relative density of at least $1/8$, it follows that $f'_{ap}(x)$ does not exist at any $x \in E$. Thus the example satisfies the properties claimed for it.

The fact that $|f(E)\cap I| \le |E\cap I|$ in the previous example suggests the characterizations of functions satisfying (N) which follow. The first of these is 'global' and the second involves the behavior at points. In each, an exceptional set whose image is of measure 0 is allowed.

THEOREM 1. A necessary and sufficient condition for a continuous function F to satisfy Lusin's condition (N) on an interval I is that $I = \bigcup_{n=0}^{\infty} E_n$, $|F(E_n)| \leq n|E_n|$ and for $n > 0$ the E_n can be chosen so that F is one to one on each E_n.

PROOF. It is obvious that the condition is sufficient. So let $F(x)$ be defined on $[0,1]$ and satisfy Lusin's condition (N). According to the Lemma in [4] p. 283 the interval $[0,1]$ is decomposable into a sequence of measurable sets Q_n on each of which F is one to one along with a set Z for which $|F(Z) = 0$. Let $h_n(x) = |F(Q_n) \cap (0,x)|$. Since $F \in (N)$, $h_n(x)$ is a nondecreasing absolutely continuous function. For each interval $J = (a,b)$

$$\begin{aligned} h_n(b) - h_n(a) &= |F(Q_n \cap (0,b))| - |F(E \cap (0,a))| \\ &= |F(Q_n \cap (a,b))| \end{aligned}$$

because F is one to one on Q_n. It follows that for each set E, $|H_n(E)| = |F(Q_n \cap E)|$. Let

$$E_{ni} = \{x \in Q_n | h'(x) \text{ exists and } h'(x) \leq i\} .$$

Then for any interval J,

$$|F(E_{ni} \cap J)| = |h_n(E_{ni} \cap J)| \leq i|E_{ni} \cap J| .$$

By adding to the set Z the set of points x in each Q_n where $h_n'(x)$ does not exist, since these form a set of measure 0, it still follows that $|F(Z)| = 0$. By rearranging the sets E_{ni} into a single sequence $\{E_m\}$ where if $E_{ni} = E_m$, $m > i$, we have $|F(E_m \cap J)| \leq m|E_m \cap J|$ for each interval J. Thus the condition is also necessary.

Theorem 1 does not give a picture of the way each $F \in (N)$ 'mixes' the measure of each set E_m. This suggests the following problem.

PROBLEM. If $F \in (N)$ does there exist a set Z with $|F(Z)| = 0$ so that $I \backslash Z$ is a countable union of sets on each of which the graph of F has finite Hausdorff length|

It is tempting to try and reduce Theorem 1 to a pointwise characterization by insisting that at each x, except on a set whose image is of measure 0, there exists E_x and n_x such that F is one to one on E_x, x is a point of density of E_x and $|F(E_x \cap I)| \leq n_x|E_x \cap I|$. However, if $s(x)$ is an increasing singular function and $f(x) = s(x) + x$, then, letting $E_x = \{x\} \cup \{t | f'(t)=1\}$.

E_x satisfies the condition stated at each x. Clearly, f does not satisfy (N). One can also observe that $f(E_x)$ does not have $f(x)$ as a point of density at each point of a set of positive measure of $y = f(x)$. When this type of behavior is disallowed, a theorem can be developed characterizing (N).

THEOREM 2. In order that a continuous function F satisfy Lusin's condition (N) on an interval I, it is necessary and sufficient that at each x, except on a set whose image is of measure 0, there exist E_x and n_x so that F is one to one on E_x, x is a point of density of E_x, $F(x)$ is a point of density of $F(E_x)$ and $|F(E_x \cap J)| \le n_x |E_x \cap J|$ for each interval J.

PROOF. Suppose the property claimed holds and (N) does not for a given $F(x)$. Then it is possible to choose a closed set E so that $|E| = 0$, F is one to one on E and $|F(E)| > 0$. Clearly, E can be chosen so that it contains no points of the exceptional set whose image is of measure 0. Consider the collection of sets

$$\{F(H_x \cap J)\}$$

where x is any point of E, J any closed interval, F is one to one on H_x, $H_x \subset E_x$, H_x is closed, and $F(H_x)$ has $F(x)$ as a point of density. This collection of sets covers $F(E)$ in the sense of Vitalli. Thus, there is an atmost countable pairwise disjoint collection

$$\{H_{x_i} \cap J_i\}$$

so that

$$|F(E) \setminus \bigcup_i F(H_{x_i} \cap J_i)| = 0 .$$

Then $\cup F(E_{x_i})$ contains almost every point of $F(E)$. That F satisfies (N) on these sets contradicts the assumption that $|F(E)| > 0$. Thus F satisfies (N).

Note that the proof in this direction does not require continuity for F; i.e., 'continuous' could be replaced by 'is a Baire Function'; cf. [1].

The other direction of the proof follows from Theorem 1. Let $M_n = \{x \in E_n | F(x)$ is a point of density of $F(E_n)\}$ where the E_n are given by Theorem 1. Then H_n is almost all of the set $F(E_n)$ and the points of $F(E_n) \setminus H_n$ can be added to the set whose image is of measure 0 along with the set of points which are not points of density of E_n. Then, for each x which is a point of density of E_n with $F(x) \in H_n$, the $E_x = E_n$ and $n_x = n$ are obtained.

It is perhaps worth while to give a descriptive statement of Theorem 2. For this purpose, let $E_{a,f,x}$ be a set such that

x is a point of density of $E_{a,f,x}$

f is a one to one on $E_{a,f,x}$

$f(x)$ is a point of density of $f(E_{a,f,x})$

$|f(E_{a,f,x} \cap J)| \leq a|E_{a,f,x} \cap J|$

for all intervals J.

Define the coefficient of expansion of f at x ($Ef(x)$) by $Ef(x) = \inf\{a \mid \text{There exists } E_{a,f,x}\}$ (∞ if there are no $E_{a,f,x}$). Then a continuous function F satisfies (N) iff at every point x except on a set whose image is of measure 0, there is a finite coefficient of expansion at x.

BIBLIOGRAPHY

1. J. Foran, "A note on Lusin's condition (N)," Fund. Math., 90 (1976), 651-661.

2. ________, "A generalization of absolute continuity," Real Analysis Exchange 5 (1979), no. 1, 82-91.

3. ________, "On extending the Lebesgue Integral," Proc. Amer. Math. Soc., 81 (1981), vol. 1, 85-88.

4. S. Saks, Theory of the Integral, 2nd Ed. revised, New York.

DEPARTMENT OF MATHEMATICS
UNIVERSITY OF MISSOURI-KANSAS CITY
KANSAS CITY, MISSOURI 64110

Contemporary Mathematics
Volume 42, 1985

CONSTRUCTION OF ABSOLUTELY CONTINUOUS AND SINGULAR FUNCTIONS THAT ARE NOWHERE OF MONOTONIC TYPE

K.M. Garg[(1)]

INTRODUCTION

Let R denote the set of real numbers and let I be any subinterval of R. Given a function $f: I \to R$ and $\alpha \in R$, we use [4] $f_{+\alpha}$ to denote the function

$$f_{+\alpha}(x) = f(x) + \alpha x, \quad x \in I,$$

and $f_{-\alpha}$ to denote the function $f_{+(-\alpha)}$.

The function f is said [4] to be of <u>monotonic type</u>, or simply MT, if there exists an $\alpha \in R$ such that the function $f_{+\alpha}$ is monotone on I. Further, f is called <u>nowhere of monotonic type</u>, or nowhere MT[(2)], if it is not MT on any (nondegenerate) subinterval of I. The function f is similarly <u>nowhere monotone</u> if it is not monotone on any subinterval of I.

In this paper we present a simple new method of construction of continuous singular functions and of absolutely continuous functions that are nowhere MT. Two such classes of functions are defined geometrically by this method and then analytical definitions are given for two particular examples of the two classes of functions. By analytical definitions we mean arithmetical definitions, for they enable a closer analysis of the properties of the functions. To demonstrate this, we include here direct proofs of the fact that the two functions are nowhere MT in terms of their analytical definitions.

There seems to be only one known example [1][(3)] of a continuous singular function in the literature which is nowhere monotone. This function

---

[1] The present work was supported in part by the NSERC of Canada Grant No. A4826.

[2] A nowhere MT function is also known [2] as a nowhere monotone function of the second species.

[3] This example is a modified version of a function constructed originally by U.K. Shukla [12] which was found in [13] to be discontinuous at a dense set of points.

0271-4132/85 $1.00 + $.25 per page

is defined analytically and, since any singular function is nowhere monotone iff it is nowhere MT (see [4, p. 311]), it is indeed nowhere MT. The present definition of such functions is more intuitive and the analytical definition is found to be simpler.

There are several known examples of absolutely continuous nowhere monotone functions. For example, the Köpcke's everywhere derivable nowhere monotone function [8, pp. 412-421] can be easily seen to be a Lipschitz function, or see Exercise 18.31 of [7, p. 296] for a simpler example. But we do not know if there is any analytically defined absolutely continuous function in the literature which is nowhere MT. The existence of functions with these properties does follow, however, from some of the known results.

Consider, e.g., the following result of Zahorski [14, p. 102]. Given any G_δ-set E of measure zero in $I = [0,1]$, there exists an absolutely continuous function f on I such that E is exactly the set of points where $D^+f = D^-f = \infty$ and $D_+f = D_-f = -\infty$. On choosing E to be dense in I it is clear that f is further nowhere MT.

Or, given any $F_{\sigma\delta}$-set E of measure zero in $I = [0,1]$, it has been proved by Landis [10] that there exists a continuous function f on I such that $E = \{x \in I : f'(x) = \infty\}$. The function constructed by him is indeed absolutely continuous. Hence, given any two disjoint countable dense sets A and B in I, there exist two absolutely continuous functions g and h on I such that $A = \{x \in I : g'(x) = \infty\}$ and $B = \{x \in I : h'(x) = \infty\}$. The function $f = g - h$ is then absolutely continuous, and it is easy to see that $\bar{D}f = \infty$ and $\underline{D}f = -\infty$ at each point of A and B respectively. Consequently, f is nowhere MT.

As regards the organization of the sections, we first discuss in §1 our method of construction. This includes some definitions and preliminary results which are basic to the method. Section 2 is devoted to the construction of a class of continuous, singular, nowhere MT functions, and in §3 we present an analytical definition of a particular function of that class. Section 4 deals similarly with the construction of a class of absolutely continuous nowhere MT functions and §5 with the analytical definition of a particular function of that class.

§1. THE METHOD OF CONSTRUCTION

From now on we shall assume, for the sake of definiteness, that $I = [0,1]$ and use $\mathcal{B}$ be denote the linear space of all real-valued functions of bounded variation on I. For each function $f \in \mathcal{B}$, we use f^+, f^- and $\bar{f}$ to denote the positive, negative and total variation functions respectively of f on I. The functions f^+ and f^- will be called simply the positive and negative variations respectively of f.

Let a function $f : I \to R$ be called further nowhere constant or nowhere Lipschitz if it is not constant or Lipschitz respectively on any subinterval of I. Our constructions are based on the following characterizations of nowhere monotone and nowhere MT functions in $\mathcal{B}$.

THEOREM 1.1. Given $f \in \mathcal{B}$,

(a) f is nowhere monotone iff both f^+ and f^- are nowhere constant, and

(b) f is nowhere MT iff both f^+ and f^- are nowhere Lipschitz.

Proof. (a) It is enough to prove, for this part, that f is monotone on some subinterval of I iff one of the functions f^+ and f^- is constant on some subinterval of I. This is, however, obvious since $f = f^+ - f^-$.

(b) It is enough to prove here that f is MT on some subinterval of I iff one of the functions f^+ and f^- is Lipschitz on some subinterval of I. To prove the necessity, suppose f is MT on some subinterval J of I. Then there exists an $\alpha \geq 0$ such that either (i) $f_{-\alpha}$ is nonincreasing on J or (ii) f_α is nondecreasing on J. Suppose first that (i) holds. Then for each $x, y \in J$, $x < y$, we have $f(y) - f(x) \leq \alpha(y-x)$. Hence if $x, y \in J$ and $x < y$, it follows easily from the definition of f^+ that

$$0 \leq f^+(y) - f^+(x) \leq \alpha(y-x).$$

Consequently, f^+ is Lipschitz on J. In the case (ii) it is proved similarly that f^- is Lipschitz on J.

To prove the sufficiency, suppose first that f^+ is Lipschitz on some subinterval J of I. Then there exists an $\alpha > 0$ such that the function $(f^+)_{-\alpha}$ is nonincreasing on J. Hence for each $x, y \in J$, $x < y$, we have

$$f(y) - f(x) \leq f^+(y) - f^+(x) \leq \alpha(y-x).$$

The function $f_{-\alpha}$ is thus nonincreasing on J, and f is consequently MT on J. A similar argument holds when f^- is Lipschitz on some subinterval of I. The proof is now complete.

Given $f \in \mathcal{B}$, let Vf denote the total variation of f on I. Then

$$\|f\| = |f(0)| + Vf, \quad f \in \mathcal{B},$$

is a norm on $\mathcal{B}$ relative to which $\mathcal{B}$ is a Banach space (see e.g. [5]).

Let $\mathcal{B}_c$, $\mathcal{B}_a$, $\mathcal{B}_s$ and $\mathcal{B}_{cs}$ denote the linear subspaces of $\mathcal{B}$ consisting of continuous, absolutely continuous, singular and continuous singular functions respectively in $\mathcal{B}$. Each of these subspaces is known to be complete (see e.g. [3] or [5]).

Let a series of functions Σf_n in $\mathcal{B}$ be called s-convergent if the two series $\Sigma f_n(0)$ and $\Sigma V f_n$ are convergent in R. Following is an extension of the Fubini theorem obtained in [5].

THEOREM 1.2. If a series of functions Σf_n in $\mathcal{B}$ is s-convergent, then it converges in the norm to a function $f \in \mathcal{B}$ and $f'(x) = \Sigma_{n=1}^{\infty} f_n'(x)$ for almost all x in I.

Given two functions $f, g \in \mathcal{B}$, we call them (see [5])

(i) mutually singular (or $f \perp g$) if for each $\varepsilon > 0$ there exists a partition $0 = x_0 < x_1 < \dots < x_n = 1$ of I with a decomposition (S_-, S_+) of $S_n \equiv \{1,2,\dots,n\}$ such that

$$\sum_{i \in S_+} \{\overline{f}(x_i) - \overline{f}(x_{i-1})\} + \sum_{i \in S_-} \{\overline{g}(x_i) - \overline{g}(x_{i-1})\} < \varepsilon ,$$

(ii) mutually lower singular (or $f \perp_- g$) if for each $\varepsilon > 0$ there exists a partition $0 = x_0 < x_1 < \dots < x_n = 1$ of I with a decomposition (S_-, S_+) of

$$S = \{i \in S_n : \big(f(x_i) - f(x_{i-1})\big)\big(g(x_i) - g(x_{i-1})\big) < 0\}$$

such that $\sum_{i \in S_n \sim S_-} |f(x_i) - f(x_{i-1})| > Vf - \varepsilon$ and $\sum_{i \in S_n \sim S_+} |g(x_i) - g(x_{i-1})| > Vg - \varepsilon$,

(iii) mutually upper singular (or $f \perp^- g$) if $f \perp_- (-g)$.

As seen in [5], $f \perp g$ iff $f \perp_- g$ and $f \perp^- g$. Further, $f \perp_- g$ iff $V(f + g) = Vf + Vg$, and similar statements hold for all the other variations. The following result holds similarly for a series of functions in $\mathcal{B}$ (see [5]).

THEOREM 1.3. Let Σf_n be a series of functions in $\mathcal{B}$ which converges in the norm to $f \in \mathcal{B}$. Then the functions in Σf_n are pairwise (mutually) lower singular iff any of the following equivalent conditions holds:

$$\text{(a)}\quad \overline{f} = \sum_{n=1}^{\infty} \overline{f}_n, \qquad \text{(b)}\quad f^+ = \sum_{n=1}^{\infty} f_n^+, \qquad \text{(c)}\quad f^- = \sum_{n=1}^{\infty} f_n^- .$$

What is needed here is indeed the following easy consequence of the above three results:

COROLLARY 1.4. Let $\{f_n\}$ be any pairwise singular sequence of nondecreasing functions on I such that $f_n(0) = 0$ for each n and the series $\Sigma f_n(1)$ is convergent. Then

$$f^+ = \sum_{n=1}^{\infty} f_{2n-1} \quad \text{and} \quad f^- = \sum_{n=1}^{\infty} f_{2n}$$

are the positive and negative variations of the function $f = f^+ - f^-$ which is of bounded variation.

Moreover, if all the functions in $\{f_n\}$ are continuous, singular or absolutely continuous, then so is f.

Given any set $E \subset R$, we use $\mathrm{co}\, E$ to denote the convex hull of E and $|E|$ to denote the Lebesgue outer measure of E. To verify when two functions are mutually singular, we further need the following result of [5].

THEOREM 1.5. Two functions $f, g \in \mathcal{B}$ are mutually singular iff the following conditions hold:

(a) f and g are not discontinuous from the same side at any point of I,

(b) $f'(x)g'(x) = 0$ for almost all x in I, and

(c) the set of points where both f and g have infinite derivatives has a decomposition into two Borel sets A,B such that $|\overline{f}(A)| = |\overline{g}(B)| = 0$.

Since every absolutely continuous function satisfies the Lusin's condition (N) (see [11, p. 225]), we obtain clearly, from the above theorem,

COROLLARY 1.6. Two functions $f, g \in \mathcal{B}$ are mutually singular whenever one of the following two conditions holds:

(a) both f and g are continuous and singular and the set of points where they both have infinite derivatives is countable,

(b) both f and g are absolutely continuous and $f'(x)g'(x) = 0$ for almost all x in I.

Finally, we call a function f on I a <u>step function</u> if it is continuous and nondecreasing and if there exists a sequence of intervals $\{I_n\}$ in I such that $\bigcup_n I_n$ is dense in I and f is constant on each I_n.

Our method of construction can now be described as follows. We first use Corollary 1.6 to construct a pairwise singular sequence of singular or absolutely continuous step functions $\{\phi_n\}$ on I such that $\phi_n(0) = 0$ for each n. Now let $\{\alpha_n\}$ and $\{\beta_n\}$ be two sequences of positive numbers such that the series $\Sigma\alpha_n\phi_{2n-1}(1)$ and $\Sigma\beta_n\phi_{2n}(1)$ are convergent. Then

$$f^+ = \sum_{n=1}^{\infty} \alpha_n\phi_{2n-1} \quad \text{and} \quad f^- = \sum_{n=1}^{\infty} \beta_n\phi_{2n}$$

are by Corollary 1.4 the positive and negative variations of the function $f = f^+ - f^-$ which in turn is continuous, singular or absolutely continuous respectively. Using this decomposition of f we then prove with the help of Theorem 1.1 that f is nowhere MT.

§2. A CLASS OF CONTINUOUS, SINGULAR, NOWHERE MT FUNCTIONS

Using the well-known Lebesgue's singular step function on I (see e.g. [7, p. 113]), we first construct a sequence of singular step functions $\{\phi_n\}$ on I.

Let P_0 denote the Cantor's nowhere dense perfect set [7, p. 70] in I. Given any closed subinterval $J = [a,b]$ of I, let $P_J \equiv P_{a,b}$ denote the image of P_0 by the linear map

$$\lambda(x) = a + (b-a)x, \quad 0 \leq x \leq 1.$$

It is easy to see that P_J is a nowhere dense perfect set such that co $P_J = J$ and $|P_J| = 0$.

Next, given any nowhere dense perfect set P in I such that co P = I, let $\{I_n\}$ be the sequence of (closed) contiguous intervals of P in I. We define two sets $P^\sim$ and P^* corresponding to P as follows:

$$P^\sim = \bigcup_{n=1}^{\infty} P_{I_n}, \quad P^* = P \cup P^\sim .$$

Then P^* is again a nowhere dense perfect set in I such that $|P^*| = 0$ and $\text{co}(P^*) = I$.

Now set $P_1 = P_0$ and define by induction, for each integer $n > 1$,

$$P_n = P^*_{n-1} .$$

Then $\{P_n\}$ is clearly a nondecreasing sequence of nowhere dense perfect sets in I such that $\cup_{n=1}^{\infty} P_n$ is a dense meager set in I whose measure is zero.

Next, let ϕ denote the Lebesgue's singular step function on I. Given any nowhere dense perfect set P of measure zero with co P = I, we define another step function ϕ_P on I as follows. Let $\phi_P(x) = x$ for each $x \in P$ and define, for each contiguous interval [a,b] of P in I,

$$\phi_P(x) = a + (b - a)\,\phi\left(\frac{x - a}{b - a}\right), \qquad a < x < b.$$

Since $\phi(0) = 0$ and $\phi(1) = 1$, it is clear that ϕ_P is a step function on I which is constant on each contiguous interval of P*. Hence ϕ_P is singular.

We now define $\phi_1 = \phi$ and $\phi_n = \phi_{P_{n-1}}$ for $n > 1$. Then ϕ_n is, for each n, a singular step function on I with $\phi_n(0) = 0$ and $\phi_n(1) = 1$ such that ϕ_n is constant on each contiguous interval on P_n.

We are now ready to construct the desired class of functions. Let $\Sigma\alpha_n$ and $\Sigma\beta_n$ be any two convergent series of positive numbers. Set, for each positive integer n,

$$f_{2n-1} = \alpha_n\phi_{2n-1} \quad \text{and} \quad f_{2n} = \beta_n\phi_{2n}.$$

Then $\{f_n\}$ is a sequence of singular step functions on I such that $f_n(0) = 0$ for each n and

$$\sum_{n=1}^{\infty} f_n(1) = \sum_{n=1}^{\infty} \alpha_n + \sum_{n=1}^{\infty} \beta_n < \infty.$$

Next set, for each n,

$$A_n = \{x \in I : f_n'(x) = \infty\} = \{x \in I : \phi_n'(x) = \infty\}.$$

Given any pair of distinct positive integers m and n, suppose $m < n$. It is then clear that $E_m \subset P_m \subset P_{n-1}$. Further, since P_{n-1} is a perfect set and $\phi_n(x) = x$ for each $x \in P_{n-1}$, we have $\underline{D}\phi_n(x) \leq 1$ for each $x \in P_{n-1}$. Consequently, $A_m \cap A_n = \phi$, and hence it follows from Corollary 1.6 that $f_m \perp f_n$.

It follows now from Corollary 1.4 that

$$f^+ = \sum_{n=1}^{\infty} f_{2n-1} \quad \text{and} \quad f^- = \sum_{n=1}^{\infty} f_{2n}$$

are two continuous singular functions which are the positive and negative variations of the continuous singular function $f = f^+ - f^-$.

It remains to show that f is nowhere MT. Since f is singular, it is indeed enough to show by Corollary 2.25 of [4] that f is nowhere monotone. Hence it is enough to show by Theorem 1.1 that each of the functions f^+ and f^- is nowhere constant. Given an open subinterval J of I, there clearly exists an integer n such that J contains some contiguous interval $[a,b]$ of P_{2n}. Then $\phi_{2n+1}(a) = a$ and $\phi_{2n+1}(b) = b$, and hence $f_{2n+1}(a) < f_{2n+1}(b)$. Since $f_i(a) \leq f_i(b)$ for all positive integers i, we have thus $f^+(a) < f^+(b)$. This proves that f^+ is nowhere constant, and it is clear that the same holds for f^-. Consequently, f is nowhere MT.

REMARK 2.1. The above arguments hold also on replacing P_0 by any other nowhere dense perfect set of measure zero with $\text{co}\, P_0 = I$, and ϕ by any continuous nondecreasing function on I which is increasing on P_0 but constant on each contiguous interval of P_0.

Further, as seen above, the functions f^+ and f^- provide new examples of the frequently considered class of continuous increasing singular functions.

§3. ANALYTICAL DEFINITION OF A CONTINUOUS, SINGULAR, NOWHERE MT FUNCTION

We obtain here analytical definitions of the functions f, f^+ and f^- of the last section with $\alpha_n = \beta_n = 2^{-n}$ for each n. Using this definition of f, we then prove directly that f is nowhere MT.

Let $A_{n,\infty}$ denote, for each positive integer n, the infinite series

$$A_{n,\infty} = \frac{a_{n,1}}{3} + \frac{a_{n,2}}{3^2} + \dots + \frac{a_{n,i}}{3^i} + \dots ,$$

where $a_{n,i} = 0$ or 2 for each positive integer i. Given any other positive integer k_n, we use further A_{n,k_n} to denote the finite sum

$$A_{n,k_n} = \frac{a_{n,1}}{3} + \frac{a_{n,2}}{3^2} + \dots + \frac{a_{n,k_n-1}}{3^{k_n-1}} + \frac{1}{3^{k_n}} ,$$

where $a_{n,i} = 0$ or 2 for each $i = 1,2,\dots,k_n-1$.

Then a point x belongs to the Cantor set P_1 iff it has a representation of the form $A_{1,\infty}$. Further, if $[a,b]$ is any closed subinterval of I, then $x \in P_{a,b}$ iff it is representable in the form

$$x = a + (b-a)\left\{\frac{a_1}{3} + \frac{a_2}{3^2} + \dots + \frac{a_i}{3^i} + \dots\right\},$$

where $a_i = 0$ or 2 for each $i = 1,2,\ldots$. Hence $x \in \tilde{P}_1$ iff it has a representation of the form

$$x = A_{1,k_1} + \frac{1}{3^{k_1}} A_{2,\infty}$$

for some positive integer k_1. In general, for each positive integer n, a point $x \in \tilde{P}_n$ iff it has a representation of the form

$$x = A_{1,k_1} + \frac{1}{3^{k_1}}\left(A_{2,k_2} + \cdots \left(\cdots + \frac{1}{3^{k_{n-1}}}\left(A_{n,k_n} + \frac{1}{3^{k_n}}\left(A_{n+1,\infty}\right)\right)\cdots\right)\right.$$

for some positive integers $k_1, k_2, \ldots, k_n$. Each point $x \in I \sim \cup_{n=1}^{\infty} P_n$ is expressible, on the other hand, in terms of infinitely many brackets as

$$x = A_{1,k_1} + \frac{1}{3^{k_1}}\left(A_{2,k_2} + \cdots \left(\cdots + \frac{1}{3^{k_{n-1}}}\left(A_{n,k_n} + \cdots \left(\cdots ,\right.\right.\right.\right.$$

where $\{k_n\}$ is some infinite sequence of positive integers.

Suppose each point $x \in I$ is expressed in one of the above two forms with finite or infinite number of brackets. Since

$$\frac{2}{3^i} = \frac{1}{3^i} + \frac{1}{3^i}\left(\frac{2}{3} + \frac{2}{3^2} + \cdots\right),$$

there do exist countably many points in $\cup_{n=1}^{\infty} P_n$ which have double representation, one with one bracket more than the other. The following definitions of all the functions yield however identical values for the two representations of such points.

The value of Lebesgue's function ϕ_1 at each point $x \in I$, expressed as above, is given by

$$\phi_1(x) = \frac{b_{1,1}}{2} + \frac{b_{1,2}}{2^2} + \cdots + \frac{b_{1,k_1-1}}{2^{k_1-1}} + \frac{1}{2^{k_1}},$$

where $b_{1,i} = \frac{1}{2} a_{1,i}$ for each positive integer $i < k_1$, and the last term $1/2^{k_1}$ does not appear in the case when $k_1 = \infty$ in the representation of x. Following this convention it is easy to see that the value of the function ϕ_n, for each positive integer n, is given by

$$\phi_n(x) = A_{1,k_1} + \frac{1}{3^{k_1}}\left(A_{2,k_2} + \cdots \left(\cdots + \frac{1}{3^{k_{n-2}}}\left(A_{n-1,k_{n-1}} + \frac{1}{3^{k_{n-1}}}\left(B_{n,k_n}\right)\right)\cdots\right)\right),$$

where

$$B_{n,k_n} = \frac{b_{n,1}}{2} + \frac{b_{n,2}}{2^2} + \dots + \frac{b_{n,k_n-1}}{2^{k_n-1}} + \frac{1}{2^{k_n}}$$

and $b_{n,i} = \frac{1}{2} a_{n,i}$ for each positive integer $i < k_n$.

Choosing $\alpha_n = \beta_n = 2^{-n}$ for each n, it is now easy to see that

$$f^+(x) = \frac{1}{2}\left(C_{1,k_1} + \frac{1}{3^{k_1}}\left(A_{2,k_2} + \frac{1}{2\cdot 3^{k_2}}\left(C_{3,k_3} + \frac{1}{3^{k_3}}\left(A_{4,k_4} + \dots\left(\dots\right.\right.\right.\right.\right. ,$$

$$f^-(x) = A_{1,k_1} + \frac{1}{2\cdot 3^{k_1}}\left(C_{2,k_2} + \frac{1}{3^{k_2}}\left(A_{3,k_3} + \frac{1}{2\cdot 3^{k_3}}\left(C_{4,k_4} + \dots\left(\dots\right.\right.\right.\right. ,$$

where, for each $n = 1,2,\dots,$

$$C_{n,k_n} = \frac{c_{n,1}}{6} + \frac{c_{n,2}}{6^2} + \dots + \frac{c_{n,k_n-1}}{6^{k_n-1}} + \frac{3^{k_n} + 2^{k_n}}{6^{k_n}}$$

and

$$\left.\begin{aligned} c_{n,i} &= 0 && \text{if } a_{n,i} = 0 \\ &= 3^i + 2^{i+1} && \text{if } a_{n,i} = 2 \end{aligned}\right\} \quad (i = 1,\dots,k_n-1)\,.$$

Thus we obtain, finally,

$$f(x) = \frac{1}{2}\left(D_{1,k_1} - \frac{1}{3^{k_1}}\left(B_{2,k_2} - \frac{1}{2\cdot 3^{k_2}}\left(D_{3,k_3} - \frac{1}{3^{k_3}}\left(B_{4,k_4} - \dots\left(\dots\right.\right.\right.\right.\right. ,$$

where, for each $n = 1,2,\dots$,

$$D_{n,k_n} = \frac{d_{n,1}}{6} + \frac{d_{n,2}}{6^2} + \dots + \frac{d_{n,k_n-1}}{6^{k_n-1}} + \frac{3^{k_n} - 2^{k_n}}{6^{k_n}}$$

and

$$\left.\begin{aligned} d_{n,i} &= 0 && \text{if } a_{n,i} = 0 \\ &= 3^i - 2^{i+1} && \text{if } a_{n,i} = 2 \end{aligned}\right\} \quad (i = 1,\dots,k_n-1).$$

Next, to see that f is nowhere MT, let J be an arbitrary open subinterval of I and α be any positive number. On choosing n large enough, there clearly exists a contiguous interval J_0 of P_{2n} such that

$J_0 \subset J$. If x is the left endpoint of J_0, it must be of the form

$$x = A_{1,k_1} + \frac{1}{3^{k_1}}\left(A_{2,k_2} + \cdots \left(\cdots + \frac{1}{3^{k_{2n-1}}}\left(A_{2n,k_{2n}}\right)\right)\cdots\right).$$

Let m be any positive integer such that $(3/2)^m > 2^{n+1}\alpha + 1$ and set

$$y = A_{1,k_1} + \frac{1}{3^{k_1}}\left(A_{2,k_2} + \cdots \left(\cdots + \frac{1}{3^{k_{2n-1}}}\left(A_{2n,k_{2n}} + \frac{1}{3^{k_{2n}}}\left(\frac{1}{3^m}\right)\right)\right)\cdots\right)$$

and

$$z = A_{1,k_1} + \frac{1}{3^{k_1}}\left(A_{2,k_2} + \cdots \left(\cdots + \frac{1}{3^{k_{2n-1}}}\left(A_{2n,k_{2n}} + \frac{1}{3^{k_{2n}}}\left(\frac{1}{3^m} + \frac{1}{3^m}\left(\frac{1}{3^m}\right)\right)\right)\cdots\right)\right).$$

Then $x < y < z$, where $x,y,z \in J_0 \subset J$. It is now easy to verify that

$$\frac{f(y)-f(x)}{y-x} = \left\{\frac{(-1)^{2n}}{2^{n+1}3^{k_1+k_2+\cdots+k_{2n}}} \cdot \frac{3^m-2^m}{6^m}\right\}\left\{\frac{1}{3^{k_1+k_2+\cdots+k_{2n}}} \cdot \frac{1}{3^m}\right\}^{-1}$$

$$= \frac{3^m-2^m}{2^{m+n+1}} = \frac{1}{2^{n+1}}\left\{\left(\frac{3}{2}\right)^m - 1\right\} > \alpha$$

and that

$$\frac{f(z)-f(y)}{z-y} = \left\{\frac{(-1)^{2n+1}}{2^{n+1}3^{k_1+k_2+\cdots+k_{2n}+m}} \cdot \frac{3^m-2^m}{6^m}\right\}\left\{\frac{1}{3^{k_1+k_2+\cdots+k_{2n}+m}} \cdot \frac{1}{3^m}\right\}^{-1}$$

$$= -\frac{3^m-2^m}{2^{m+n+1}} = -\frac{1}{2^{n+1}}\left\{\left(\frac{3}{2}\right)^m - 1\right\} < -\alpha .$$

Hence $f_{-\alpha}(y) > f_{-\alpha}(x)$ and $f_{+\alpha}(z) < f_{+\alpha}(y)$. Thus neither the function $f_{+\alpha}$ is nondecreasing on J nor $f_{-\alpha}$ is nonincreasing on J. Consequently, f is nowhere MT.

§4. A CLASS OF ABSOLUTELY CONTINUOUS NOWHERE MT FUNCTIONS

We begin with the construction of a sequence of absolutely continuous step functions $\{\Psi_n\}$ on I.

Let P_0 be this time a nowhere dense perfect set such that $\text{co } P_0 = I$ and $\theta \equiv |P_0| > 0$. Given any closed subinterval $J = [a,b]$ of I, let $P_J \equiv P_{a,b}$ denote as before the image of P by the linear map

$$\lambda(x) = a + (b-a)x, \quad 0 \leq x \leq 1.$$

It is easy to see that P_J is a nowhere dense perfect set such that $\mathrm{co}\, P_J = J$ and $|P_J| = \theta(b-a)$.

Next, given any nowhere dense perfect set P with $\mathrm{co}\, P = I$, let $\{I_n\}$ be the sequence of contiguous intervals of P in I. The sets $P^\sim$ and P^* corresponding to P are defined as before by

$$P^\sim = \bigcup_{n=1}^{\infty} P_{I_n}, \qquad P^* = P \cup P^\sim .$$

Then $P^\sim$ is an F_σ-set such that $P \cap P^\sim$ is countable and $|P^\sim \cap I_n| > 0$ for each n. The set P^* is again a nowhere dense perfect set with $\mathrm{co}\, P^* = I$. It is further clear that

$$(1) \qquad |P^\sim| = \theta(1 - |P|), \qquad |P^*| = \theta + (1-\theta)|P| .$$

Now set $P_1 = P_0$ and define by induction, for each $n > 1$,

$$(2) \qquad P_n = P^*_{n-1} .$$

Then $\{P_n\}$ is a nondecreasing sequence of nowhere dense perfect sets in I. Set, further,

$$(3) \qquad E_1 = P_0 \quad \text{and} \quad E_n = P^\sim_{n-1} \quad \text{for} \quad n = 2,3,\ldots .$$

Then $\{E_n\}$ is a sequence of nowhere dense F_σ-sets in I no two of which have more than countably many points in common. Clearly $|E_1| = \theta$, and for each $n > 1$ we have by (1) and (2),

$$|P_n| = \theta + (1-\theta)|P_{n-1}| .$$

With the help of this recurrence relation, we obtain from (1) and (3),

$$\begin{aligned} |E_n| = |P^\sim_{n-1}| &= \theta(1 - |P_{n-1}|) = \theta\{1 - \theta - (1-\theta)|P_{n-2}|\} \\ &= \theta(1-\theta)(1 - |P_{n-2}|) = \theta(1-\theta)^2(1 - |P_{n-3}|) = \cdots \\ &= \theta(1-\theta)^{n-2}(1 - |P_1|) = \theta(1-\theta)^{n-1} . \end{aligned}$$

The set $I \sim \bigcup_{n=1}^{\infty} E_n$ is consequently a residual set of measure zero.

We now define, for each n,

$$(4) \qquad \Psi_n(x) = |E_n \cap (0,x)|, \quad 0 \le x \le 1 .$$

Then Ψ_n is clearly a nondecreasing Lipschitz function on I which is constant on each contiguous interval of P_n. Further, $\Psi_n(0) = 0$ for each n, $\Psi_1(1) = \theta$ and $\Psi_n(1) = \theta(1-\theta)^{n-1}$ for $n > 1$.

We are now ready to construct the desired class of functions. Let $\{\alpha_n\}$ and $\{\beta_n\}$ be any two unbounded sequences of positive numbers such that each of the series $\Sigma\alpha_n(1-\theta)^{2n}$ and $\Sigma\beta_n(1-\theta)^{2n}$ is convergent. Set, for each positive integer n,

$$(5) \qquad g_{2n-1} = \alpha_n \Psi_{2n-1} \quad \text{and} \quad g_{2n} = \beta_n \Psi_{2n}.$$

Then $\{g_n\}$ is a sequence of nondecreasing Lipschitz functions on I such that $g_n(0) = 0$ for each n and

$$\sum_{n=1}^{\infty} g_n(1) = \sum_{n=1}^{\infty} \alpha_n \Psi_{2n-1}(1) + \sum_{n=1}^{\infty} \beta_n \Psi_{2n}(1)$$

$$= \alpha_1\theta + \frac{\theta}{(1-\theta)^2} \sum_{n=2}^{\infty} \alpha_n (1-\theta)^{2n} + \frac{\theta}{1-\theta} \sum_{n=1}^{\infty} \beta_n (1-\theta)^{2n} < \infty.$$

Further, for each n, Ψ_n is clearly the indefinite integral of the characteristic function χ_{E_n} of E_n, and hence, by the Lebesgue theorem, $\Psi_n'(x) = \chi_{E_n}(x)$ for almost all x in I. Given any pair of distinct positive integers m and n, since $E_m \cap E_n$ is countable, we have thus $\Psi_m'(x)\Psi_n'(x) = 0$ for almost all x in I. Hence $g_m'(x)g_n'(x) = 0$ for almost all x in I. Consequently, $g_m \perp g_n$ by Corollary 1.6.

It follows now from Corollary 1.4 that

$$(6) \qquad g^+ = \sum_{n=1}^{\infty} g_{2n-1} \quad \text{and} \quad g^- = \sum_{n=1}^{\infty} g_{2n}$$

are two absolutely continuous functions which are the positive and negative variations of the absolutely continuous function $g = g^+ - g^-$.

To see that g is nowhere MT, it suffices to show by Theorem 1.1 that each of the functions g^+ and g^- is nowhere Lipschitz. Let J be an arbitrary open subinterval of I and α be any positive number. Clearly, there exists an integer k such that J contains a contiguous interval J_0 of P_k. Now, as the sequence $\{\alpha_n\}$ is unbounded, there exists an integer $n > k+2$ such that $\alpha_n > \alpha$. Since $P_k \subset P_{2n-2}$, J_0 contains some contiguous interval J_1 of P_{2n-2}, and hence according to the definition of $\tilde{P}_{2n-2}$ we have

$$|J_1 \cap E_{2n-1}| = |J_1 \cap \tilde{P}_{2n-2}| > 0.$$

It follows now from (4), as earlier, that J_1 contains some point x such that $\Psi'_{2n-1}(x) = 1$. Thus $g'_{2n-1}(x) = \alpha_n > \alpha$. Since the function

$$h = \sum_{i=1}^{n-1} g_{2i-1} + \sum_{i=n+1}^{\infty} g_{2i-1}$$

is nondecreasing, $\underline{D}h(x) \geq 0$, and since $g^+ = g_{2n-1} + h$, we have

$$\underline{D}g^+(x) = g'_{2n-1}(x) + \underline{D}h(x) > \alpha .$$

This proves that g^+ is nowhere Lipschitz. It is proved similarly that g^- is nowhere Lipschitz. Consequently, g is nowhere MT.

REMARK 4.1. If we choose $\theta = 1/2$, as we do in the next section, the calculations in the above construction become considerably simpler. For then it is trivial to see that $|E_n| = 2^{-n}$ for each n.

§5. ANALYTICAL DEFINITION OF AN ABSOLUTELY CONTINUOUS NOWHERE MT FUNCTION

We obtain here analytical definitions of the functions g, g^+ and g^- of the last section with $\theta = 1/2$, $\alpha_n = 2^{n-1}$ and $\beta_n = 2^n$ for each n. It is clear that the two sequences $\{\alpha_n\}$ and $\{\beta_n\}$ have the required properties. Using this definition of g, we prove again directly that g is nowhere MT.

We need to define first a nowhere dense perfect set P_0 in I with $\text{co}\, P_0 = I$ such that $|P_0| = 1/2$. Delete from I its concentric open subinterval of length $1/4$; from each of the remaining two closed intervals delete its concentric open subinterval of length $1/4^2$; from each of the remaining four closed intervals delete its concentric open subinterval of length $1/4^3$; and so on. Let P_0 be the set of points that are left in I after continuing this process indefinitely. Since the total length of the disjoint family of deleted open intervals is $\Sigma_{n=1}^{\infty} 2^{n-1}/4^n = 1/2$, it is clear that P_0 has all the required properties.

Let $A_{n,\infty}$ denote this time, for each positive integer n, the convergent series

$$A_{n,\infty} = \frac{a_{n,1}}{4} + \frac{a_{n,2}}{4^2} + \dots + \frac{a_{n,i}}{4^i} + \dots ,$$

where $a_{n,i} = 0$ or $2^i + 3$ for each positive integer i. Given any other positive integer k_n, let us use A_{n,k_n} to denote the finite sum

$$A_{n,k_n} = \frac{a_{n,1}}{4} + \frac{a_{n,2}}{4^2} + \dots + \frac{a_{n,k_n-1}}{4^{k_n-1}} + \frac{2^{k_n}+1}{4^{k_n}},$$

where $a_{n,i} = 0$ or $2^i + 3$ for each $i = 1,2,\dots,k_n-1$.

It is easy to verify that a point $x \in P_0$ iff it is representable in the form $\frac{1}{2} A_{1,\infty}$. Further, if $[a,b]$ is any closed subinterval of I, then $x \in P_{a,b}$ iff it is representable in the form

$$x = a + \frac{b-a}{2}\left\{\frac{a_1}{4} + \frac{a_2}{4^2} + \dots + \frac{a_i}{4^i} + \dots\right\},$$

where $a_i = 0$ or $2^i + 3$ for each i. Using this fact repeatedly, it is now easy to see that, for each positive integer n, a point $x \in E_n$ iff it is representable in the form

$$x = \frac{1}{2}\left(A_{1,k_1} + \frac{1}{4^{k_1}}\left(A_{2,k_2} + \dots\left(\dots + \frac{1}{4^{k_{n-2}}}\left(A_{n-1,k_{n-1}} + \frac{1}{4^{k_{n-1}}}\left(A_{n,\infty}\right)\right)\dots\right)\right)\right)$$

for some positive integers $k_1,k_2,\dots,k_{n-1}$. Each point $x \in I \sim \bigcup_{n=1}^{\infty} E_n$ is representable, on the other hand, in terms of infinitely many brackets as

$$x = \frac{1}{2}\left(A_{1,k_1} + \frac{1}{4^{k_1}}\left(A_{2,k_2} + \dots\left(\dots + \frac{1}{4^{k_{n-1}}}\left(A_{n,k_n} + \dots\left(\dots,\right.\right.\right.\right.\right.$$

where $\{k_n\}$ is some infinite sequence of positive integers.

Suppose each point $x \in I$ is expressed in one of the above two forms. Since

$$\frac{2^i+3}{4^i} = \frac{2^i+1}{4^i} + \frac{1}{4^i}\left(\frac{2+3}{4} + \frac{2^2+3}{4^2} + \dots\right),$$

there exist as before countably many points in $\bigcup_{n=1}^{\infty} E_n$ which have double representation. The following functions assume once again identical values for the two representations of such points.

The value of Ψ_1 at each point $x \in I$, expressed as above, is found this time to be

$$\Psi_1(x) = \frac{1}{2}\left(\frac{b_{1,1}}{2} + \frac{b_{1,2}}{2^2} + \dots + \frac{b_{1,k_1-1}}{2^{k_1-1}} + \frac{1}{2^{k_1}}\right),$$

where, for each positive integer $i < k_1$, $b_{1,i} = 0$ or 1 according as $a_{1,i} = 0$ or $2^i + 3$, and the last term $1/2^{k_1}$ does not appear in the case when $k_1 = \infty$ in the representation of x. Following this convention it is easy to see that the value of Ψ_n, for each positive integer n, is given by

$$\Psi_n(x) = \frac{1}{2^n}\left(C_{1,k_1} + \frac{2}{4^{k_1}}\left(C_{2,k_2} + \cdots\left(\cdots + \frac{2}{4^{k_{n-2}}}\left(C_{n-1,k_{n-1}} + \frac{2}{4^{k_{n-1}}}\left(B_{n,k_n}\right)\right)\cdots\right)\right)\right),$$

where

$$B_{n,k_n} = \frac{b_{n,1}}{2} + \frac{b_{n,2}}{2^2} + \cdots + \frac{b_{n,k_n-1}}{2^{k_n-1}} + \frac{1}{2^{k_n}},$$

$$\left.\begin{aligned} b_{n,i} &= 0 \text{ if } a_{n,i} = 0 \\ &= 1 \text{ if } a_{n,i} = 2^i + 3 \end{aligned}\right\} \quad (i = 1,2,\ldots,k_n - 1),$$

and for each positive integer $r < n$,

$$C_{r,k_r} = \frac{c_{r,1}}{4} + \frac{c_{r,2}}{4^2} + \cdots + \frac{c_{r,k_r-1}}{4^{k_r-1}} + \frac{1}{4^{k_r}},$$

$$\left.\begin{aligned} c_{r,i} &= 0 \text{ if } a_{r,i} = 0 \\ &= 3 \text{ if } a_{r,i} = 2^i + 3 \end{aligned}\right\} \quad (i = 1,2,\ldots,k_r - 1).$$

On choosing $\alpha_n = 2^{n-1}$ and $\beta_n = 2^n$ for each n, it is now easy to see that

$$g^+(x) = \frac{1}{2}\left(A_{1,k_1} + \frac{2}{4^{k_1}}\left(C_{2,k_2} + \frac{1}{4^{k_2}}\left(A_{3,k_3} + \frac{2}{4^{k_3}}\left(C_{4,k_4} + \cdots\left(\cdots\right.\right.\right.\right.\right.$$

and

$$g^-(x) = C_{1,k_1} + \frac{1}{4^{k_1}}\left(A_{2,k_2} + \frac{2}{4^{k_2}}\left(C_{3,k_3} + \frac{1}{4^{k_3}}\left(A_{4,k_4} + \cdots\left(\cdots\right.\right.\right.\right. .$$

Thus we obtain, finally,

$$g(x) = \frac{1}{2}\left(D_{1,k_1} - \frac{2}{4^{k_1}}\left(D_{2,k_2} - \frac{1}{4^{k_2}}\left(D_{3,k_3} - \frac{2}{4^{k_3}}\left(D_{4,k_4} - \cdots\left(\cdots\right.\right.\right.\right.\right. ,$$

where, for each n,

$$D_{n,k_n} = \frac{d_{n,1}}{4} + \frac{d_{n,2}}{4^2} + \ldots + \frac{d_{n,k_n-1}}{4^{k_n-1}} + \frac{d_{n,k_n}}{4^{k_n}},$$

$$\left.\begin{aligned} d_{n,i} &= 0 && \text{if } a_{n,i} = 0 \\ &= 2^i - 3 && \text{if } a_{n,i} = 2^i + 3 \text{ and } n \text{ is odd} \\ &= 2^i && \text{if } a_{n,i} = 2^i + 3 \text{ and } n \text{ is even} \end{aligned}\right\} \quad (i < k_n)$$

and

$$\left.\begin{aligned} d_{n,k_n} &= 2^{k_n} - 1 && \text{if } n \text{ is odd} \\ &= 2^{k_n} && \text{if } n \text{ is even} \end{aligned}\right\}.$$

Next, to see that g is nowhere MT, let J be an arbitrary open subinterval of I and α be any positive number. Clearly, there exists a positive integer n such that $2^n > 3\alpha$ and such that J contains some contiguous interval J_0 of P_{2n}. If x denotes the left endpoint of J_0, it must be of the form

$$x = \frac{1}{2}\left(A_{1,k_1} + \frac{1}{4^{k_1}}\left(A_{2,k_2} + \ldots\left(\ldots + \frac{1}{4^{k_{2n-1}}}\left(A_{2n,k_{2n}}\right)\right)\ldots\right)\right).$$

Let, further,

$$y = \frac{1}{2}\left(A_{1,k_1} + \frac{1}{4^{k_1}}\left(A_{2,k_2} + \ldots\left(\ldots + \frac{1}{4^{k_{2n-1}}}\left(A_{2n,k_{2n}} + \frac{1}{4^{k_{2n}}}\left(\frac{2+1}{4}\right)\right)\right)\ldots\right)\right)$$

and

$$z = \frac{1}{2}\left(A_{1,k_1} + \frac{1}{4^{k_1}}\left(A_{2,k_2} + \ldots\left(\ldots + \frac{1}{4^{k_{2n-1}}}\left(A_{2n,k_{2n}} + \frac{1}{4^{k_{2n}}}\left(\frac{2+1}{4} + \frac{1}{4}\left(\frac{2+1}{4}\right)\right)\right)\ldots\right)\right)\right).$$

Then $x < y < z$, where $x,y,z \in J_0 \subset J$. It is now easy to verify that

$$\begin{aligned} \frac{g(y) - g(x)}{y - x} &= \left\{\frac{1}{2}\,\frac{(-1)^{2n}2^n}{4^{k_1+k_2+\ldots+k_{2n}}} \cdot \frac{2-1}{4}\right\}\left\{\frac{1}{2}\,\frac{1}{4^{k_1+k_2+\ldots+k_{2n}}} \cdot \frac{2+1}{4}\right\}^{-1} \\ &= 2^n/3 > \alpha \end{aligned}$$

and that

$$\frac{g(z)-g(y)}{z-y} = \left\{\frac{1}{2}\,\frac{(-1)^{2n+1}2^{n+1}}{4^{k_1+k_2+\ldots+k_{2n}+1}}\cdot\frac{2}{4}\right\}\left\{\frac{1}{2}\,\frac{1}{4^{k_1+k_2+\ldots+k_{2n}+1}}\cdot\frac{2+1}{4}\right\}^{-1}$$

$$= -2^{n+2}/3 < -\alpha .$$

Hence $g_{-\alpha}(y) > g_{-\alpha}(x)$ and $g_{\alpha}(z) < g_{\alpha}(y)$. This proves as before that g is nowhere MT.

REMARK 5.1. As it is seen subsequently in [6], the nowhere MT functions form a residual set in each of the spaces $\mathcal{B}$, $\mathcal{B}_c$, $\mathcal{B}_a$, $\mathcal{B}_s$ and $\mathcal{B}_{cs}$. Further properties of these residual sets of functions are investigated in [6]. For example, $\mathcal{B}_{cs}$ contains a residual set of functions which are not monotone on any set of positive outer measure.

REFERENCES

1. K.M. Garg, "On a function of non-symmetrical differentiability", Ganita 9 (1958), 65-75. MR 21 # 5703.

2. __________, "On nowhere monotone functions. III. Functions of first and second species", Rev. Math. Pures Appl. 8 (1963), 83-90. MR 27 # 1545.

3. __________, "Characterizations of absolutely continuous and singular functions", Proc. of the Conference on the Constructive Theory of Functions (Approximation Theory) (Budapest, 1969), 183-188 (1972). MR 55 #588.

4. __________, "On bilateral derivates and the derivative", Trans. Amer. Math. Soc. 210 (1975), 295-329. MR 51 #5861.

5. __________, "The mutual singularity and relative absolute continuity of functions of bounded variation in terms of derivative", to appear.

6. __________, "Properties of residual sets of functions in BV and in some of its common subspaces", to appear.

7. E. Hewitt and K. Stromberg, Real and abstract analysis, Springer-Verlag, New York, 1965. MR 32 # 5826.

8. E.W. Hobson, The theory of functions of a real variable and the theory of Fourier's series, Vol. II, Cambridge, 2nd ed., 1926.

9. H. Kober, "On decompositions and transformations of functions of bounded variation", Ann. Math. (2) 53 (1951), 565-580. MR 12-686.

10. E.M. Landis, "On the set of points of existence of an infinite derivative", (Russian), Dokl. Akad. Nauk SSSR (N.S.) 107 (1956), 202-204. MR 17-1190.

11. S. Saks, Theory of the integral, Monografie Mat., Vol. 7, PWN, Warsaw, 1937.

12. U.K. Shukla, "On points of non-symmetrical differentiability of a continuous function III", Ganita 8 (1957), 81-104. MR 20 # 6497.

13. U.K. Skukla and K.M. Garg, "A note on a function of non-symmetrical differentiability", Ganita 9 (1958), 27-32. MR 21 # 2714.

14. Z. Zahorski, "Sur les ensembles des points de divergence de certaines intégrales singulières", Ann. Soc. Polon. Math. 19 (1946), 66-105. MR 9-89.

DEPARTMENT OF MATHEMATICS
UNIVERSITY OF ALBERTA
EDMONTON, ALBERTA
CANADA T6G 2G1

Contemporary Mathematics
Volume 42, 1985

A POROSITY CHARACTERIZATION OF SYMMETRIC PERFECT SETS

P.D. Humke and B.S. Thomson

The purpose of this paper is to show that there is a relatively large class of mathematically important sets which are measure zero and first category, but not σ-porous. Although Zajicek [Z] has provided an example of such a set, his construction is cleverly tied to the proof that the set constructed is indeed not σ-porous. Thus, although he answered an important question, his construction is so rigid as to isolate his example. In this paper then, we show that an entire gaggle of these sets is already well entrenched in the literature.

The notion of porosity was first introduced by E.P. Dolsenko [D] when investigating the boundary behavior of certain complex valued functions. The notion has been adopted by real analysts and has played a significant role in sharpening our understanding of exceptional behavior of functions. If E is a set in $\mathbb{R}$, the <u>porosity of E at the point</u> $x \in \mathbb{R}$ is defined as

$$\limsup_{r\to 0} p(E,x,r)/r$$

where $p(E,x,r)$ denotes the length of the longest open interval in $(x-r,x+r)\cap(\mathbb{R}-E)$. The set E is called <u>porous</u> if it has positive porosity at each of its points, and it is called <u>σ-porous</u> if it is a countable union of porous sets. It is evident that σ-porous sets are both of the first Baire category and of zero measure. However the converse is not true. (See [Z] and THEOREM HT below.)

If $0 < \xi < 1/2$ and $[a,b]$ is an interval, the number ξ can be used to delineate two subintervals of $[a,b]$ by extracting an ξth portion of $[a,b]$ from either end of $[a,b]$. In specific, the two delineated intervals are $[a,a+\xi(b-a)]$ and $[b-\xi(b-a),b]$. Each of these portions is an ξth portion of the original, and the nondelineated central open interval is $(1-2\xi)$th portion of the original. A sequence $\{\xi_n\}$ of such numbers can then be used to uniquely define a nowhere dense perfect subset of $[0,1]$ in the following manner. First use ξ_1 to delineate two subintervals of $[0,1]$. Subsequently, use ξ_2 to delineate two subintervals of each of the intervals found using ξ_1, and so on. The nowhere dense perfect set which is the set of all points lying in

---

1980 Mathematics Subject Classification. 54H05

infinitely many delineated intervals is called the symmetric set generated by the sequence $\{\xi_n\}$ and is denoted, $C(\xi_n)$ (or simply C if the $\{\xi_n\}$ are prominent).

Our theorem is then:

THEOREM HT. The symmetric set $C(\xi_n)$ is non σ-porous if and only if $\{\xi_n\}\to 1/2$.

Prior to proving this theorem we need to establish some notation and terminology.

Let $C(\xi_n)$ be a symmetric set. Then $C(\xi_n)$ is constructed in "stages," and the portions of those stages are important to what follows. At stage 1, for example, the unit interval has been divided into three disjoint intervals: a centered open interval of length $1-2\xi_1$ which is termed a contiguous interval of the first stage and denoted I_1^1 and two end intervals called the noncontiguous intervals of the first stage. Each of these intervals has length ξ_1 and these are denoted by J_1^1 and J_1^2 where J_1^2 lies to the right of J_1^1. Inductively, then, suppose that the closed intervals J_n^k have been defined for $k = 1,1,\ldots,2^n$ and $n<N$, and the open intervals I_n^k have been defined for $k = 1,2,\ldots,2^{n-1}$ and $n<N$ in such a way that if $i<k$, then J_n^i [respectively I_n^i] lies to the left of J_n^j [respectively I_n^j].

Now, partition each interval $J_{N-1}^k (k=1,2,\ldots,2^{N-1})$ into three intervals, the end two of which are closed with length $|J_{N-1}^k|\ \xi_N$ and the center of which is open and of length $|J_{N-1}^k|(1-2\xi_N)$.

We denote the end intervals by J_N^{k-1} and $J_{N_k}^k$ where the superscript infers left to right order, and the center interval by I_N^k. The intervals J_N^i are called the noncontiguous intervals of the Nth stage and the intervals I_N^i are called the contiguous intervals of the nth stage.

Note that each noncontiguous interval of the nth stage has length $\prod_{i=1}^n \xi_i$ and so $C(\xi_n)$ has positive measure iff

$$\lim_{n\to\infty} 2^n \prod_{i=1}^n \xi_i \neq 0 .$$

In particular, this can happen only if $\{\xi_n\}\to 1/2$, but this condition is not sufficient to insure that $C(\xi_n)$ has positive measure as the sequence $\{n/2(n+2)\}$ shows. The purpose of this note is to show that the condition $\{\xi_n\}\to 1/2$ is necessary and sufficient for $C(\xi_n)$ to be non σ-porous. Prior to proving this result, however, we define a process used to create certain closed subsets of a given symmetric perfect set.

If J is an interval and $m\in N$, we define $m*J$ to be that interval of length $m|J|$ which is concentric to J. If $\mathcal{J}$ is a set of intervals, $m*\mathcal{J} = \bigcup_{J\in\mathcal{J}} m*J$. Now, if $C = C(\xi_n)$ is a symmetric perfect set, and m, n_1, $n_2 \in N$, we define

$$m*C[n_1,n_2] = [0,1] - 1*\{I_n^i : n<n_1, i<2^{n-1}\} - m*\{I_n^i : n_1\le n\le n_2, i\le 2^{n-1}\},$$

and also,

$$m^*C[n_1] = [0,1] - 1^*\{I_n^i : n<n_1, i\leq 2^{n-1}\} - m^*\{I_n^i : n_1\leq n, i\leq 2^{n-1}\} .$$

The first of these two sets defines a finite set of disjoint closed intervals (or ϕ) while the second set is a perfect subset of C (or ϕ) which has the property that no point of it is a point of $1/m$ porosity of C. We are now in a position to prove our theorem.

THEOREM HT. <u>The symmetric set</u> $C(\xi_n)$ <u>is non</u> σ-<u>porous if and only if</u> $\{\xi_n\}\to 1/2$.

Proof. Suppose, for the necessity, that a subsequence $\{\xi_{n(k)}\}\to L < 1/2$, and let $x\in C(\xi_n)$. As $x\in C$, x is in exactly one noncontiguous interval $J_n(x)$ at each stage. Further, if $I_n(x)$ is the contiguous interval bordering $J_n(x)$, then $|I_n(x)|/|J_{n-1}(x)| = 1-2\xi_n$. Now, if we let $h(k) = |J_{n(k)-1}(x)|$, then $p(C,x,h(k)) > (1-2\xi_{n(K)})/2$ and so

$$\limsup_{h\to 0} p(C,x,h) \geq (1-2L)/2 > 0$$

Hence, C is porous at each of its points.

The more difficult part is, of course, the sufficiency. Suppose then that $\{\xi_n)\to 1/2$ and that $C = C(\xi_n)$ is σ-porous. Then C can be written as $C = \bigcup_{n=1}^{\infty} E_n$ where E_n is porous with uniform porosity, say $1/p(n) > 0$ $(p(n)\varepsilon N)$, at each of its points. We also suppose that $\{p(n)\}$ is increasing. That is, if $x\varepsilon E_n$, then $p(E_n,x) > 1/p(n)$. Now, for every n there is a $k'(n)$ such that if $k \geq k'(n)$, then

$$|1-2\xi_k| < \min\{1/\prod_{i<n} p(i) , 1/n+2\}/8 .$$

We also assume $\{k(n)\}$ is increasing. We now define a nested sequence, $\{F_n\}$, of closed subsets of C in such a way that $F_n\cap E_n = \phi$. As $\cap F_n \neq \phi$, this will contradict the fact that $C = \cup E_n$. Each F_n will be a portion of an expanded version of C . In specific,

$$(*) \qquad F_n = \left[\prod_{i<n} p(i)*C[k(n)]\right] \cap I(n)$$

where the $k(n)$ is that stage where expansion commences, and $I(n)$ is an interval, and both of these last quantities depend on the nature of E_n and its relationship with previously defined F_i . In specific:

(1) a. If $cl(E_1) \neq C$, then there is a noncontiguous interval $I(1)$ of C of a stage $k(1) > k'(1)$ such that $I(1)\cap cl(E_1) = \phi$. Let F_1 be defined then by $(*)$.

b. If $cl(E_1) = C$, we let $k(1) = k'(1)$ and let $I(1)$ be a noncontigous interval of the $k(1)$ stage. Again, F_1 is defined by $(*)$. Note that in either case, $F_1\cap E_1 = \phi$.

Assume that the increasing sequence $\{k(i):i<n\}$ has been defined and the nested sequence of intervals $\{I(i):i<n\}$ has also been defined. Further assume that corresponding closed sets given by (*) are defined in such a way that $F_i \cap E_i = \phi$ for $i < n$.

(n) a. Suppose $cl(E_n) \neq F_n$. Then there is a portion of F_n which misses $cl(E_n)$ and so there is a $k(n) > \max\{k'(n),k(n-1)\}$ such that one of the component intervals of

$$\prod_{i<n} p(i)*C[k'(n-1),k(n)]\cap I(n-1)$$

misses $cl(E_n)$. Let $I(n)$ be one such component interval, and define F_n via (*).

b. Suppose E_n is dense in F_{n-1}. In this case we take $k(n) = \max\{k(n-1)+1, k'(n)\}$ and let $I(n)$ be any component interval of

$$\prod_{i<n} p(i)*C[k(n-1),k(n)]\cap I(n-1) .$$

Again, F_n is defined by (*).

In either case $F_n \subset F_{n-1}$ because more has been deleted to obtain F_n. Also, in case a., it is evident that $F_n \cap E_n = \phi$ while in case b., it is perhaps not as patent. However, as E_n is dense in F_{n-1}, it follows that every point of E_n is a point of $1/p(n)$ porosity of F_{n-1}. This entails that every point of E_n is a point of $1/\prod_{i<n} p(i)$ porosity of the original symmetric perfect set C, and consequently, the definition of F_n precludes its intersection with E_n.

THEOREM HT also has a formulation in terms of the so called "α-sequence" associated with the defining sequence $\{\xi_n\}$. In specific, if $0 < \xi_n < 1/2$ for each n, define $\alpha_n = 1-2\xi_n$. The following theorem contains THEOREM HT (part ii.) and the aforementioned measure theoretic result (part i.) from the perspective of the sequence α_n.

THEOREM TH. <u>The symmetric Cantor set</u> $C(\xi_n)$ <u>is</u>:

i. <u>not of measure zero iff</u> $\alpha_n \in \ell_1$.

ii. <u>not σ-porous iff limit</u> $\alpha_n = 0$.

From the perspective of the α_n then, the distinction between first category-measure zero and σ-porous symmetric perfect sets is characterized by the distinction between summable sequences and those which merely have limit zero. This distinction is, perhaps, subtle, but at the same time it is quite important. In one sense THEOREM TH is not complete, for there are similar collections of symmetric perfect sets which have proven useful but are not mentioned in the theorem. One such collection of sets can be characterized as follows.

Suppose that $\alpha_n \in \ell_1$ and let $\sum_{n=1}^{N} \alpha_n = s_N$ and $\sum_{n=1}^{\infty} \alpha_n = s$. Then a set is of class S_1 iff $\{s-s_n\} \in \ell_1$. This class has been used by the first author to answer a question of M. Jodeit concerning the Hilbert Transform; the results of this investigation will be published elsewhere.

BIBLIOGRAPHY

[D] Dolzenko, E.P., "Boundary properties of arbitrary functions," Izv. Akad. Nauk SSSR, 31, pp. 3-14 (1967). English translation: Math. USSR Izv. 1, pp. 1-12 (1967).

[FH] J. Foran and P.D. Humke, "Some set theoretic properties of σ-porous sets," Real Anal. Exchange, Vol. 6, No. 1 (1981) pp. 114-119.

[Z] Zajicek, L., "Sets of σ-porosity and sets of σ-porosity (q)," Casopis Pest. Mat., 101, pp. 350-359 (1976).

DEPARTMENT OF MATHEMATICS
ST. OLAF COLLEGE
NORTHFIELD, MINNESOTA 55057

MATHEMATICS DEPARTMENT
SIMON FRASER UNIVERSITY
BURNABY, B.C.
CANADA V5A 1S6

Contemporary Mathematics
Volume 42, 1985

A METHOD FOR SHOWING GENERALIZED DERIVATIVES ARE IN BAIRE CLASS ONE

Lee Larson

ABSTRACT. A theorem is presented showing that a large class of generalized derivatives are in Baire class one. In addition, it is shown that the corresponding upper and lower derivates are in Baire class three. Applications include symmetric derivatives, approximate continuity and a convexity theorem.

§1. During the past half-century there has been a steady proliferation of generalizations to the ordinary derivative. Probably because most of these derivatives were studied as tools to be used in solving specific problems, their properties were often developed only in the context of the individual derivative being studied, and not with a view toward handling entire classes of generalized derivatives at once. For example, proofs that various specific generalized derivatives are in Baire class one are presented by Filipczak [F], Larson [L1] [L2], Tolstov [T], Preiss [P] and Zahorski [Z1]. The purpose of this paper is to present two theorems which show that a large class of generalized derivatives are in Baire class one and that the corresponding upper and lower derivates are in Baire class three. As a consequence, it is shown that symmetric derivatives, Riemann derivatives and approximately symmetric functions are in Baire class one. A convexity theorem is also proved.

In the following, the real numbers are denoted by $\mathbb{R}$ and the positive integers by $\mathbb{N}$.

If $A \subset \mathbb{R}$, then its complement with respect to $\mathbb{R}$ is A^c and if it is Lebesgue measurable, then its measure is written $|A|$. The translation of A by $t \in \mathbb{R}$ is written $A+t$; i.e., $A+t = \{x+t : x \in A\}$. The difference A and another set B is written $A-B = A \cap B^c$. The symmetric difference of A and B is denoted $A \Delta B = (A-B) \cup (B-A)$.

1980 Mathematics Subject Classification. 26A24, 26A21.

A function h is a parameter function iff it satisfies:

(1.1) h is measurable;

(1.2) if G is any measurable set, then $h(G) = \{h(x) : x \in G\}$ is measurable; and,

(1.3) if A_n is any sequence of Borel sets such that $|A_n| \to 0$ as $n \to \infty$, then

$$|h^{-1}(A_n)| \to 0 .$$

Unless specifically noted otherwise, all functions are assumed to be real-valued, measurable and have domain $\mathbb{R}$.

The approximate upper (lower) limit of a function f at x is written

$$\text{app lim sup}_{t\to x} f(t) \quad (\text{app lim inf}_{t\to x} f(t)) .$$

When these two are equal, the approximate limit of f at x, $\text{app lim}_{t\to x} f(t)$, exists.

To define the class of generalized derivatives with which this paper is concerned, let $h_1,\ldots,h_n$ be parameter functions and $a_i \in \mathbb{R}$, $i = 1,\ldots,n$. A generalized difference quotient is defined as

(1.4) $$Q(x,t) = \frac{1}{g(t)} \Sigma_{i=1}^{n} a_i f(x+h_i(t)) ,$$

where f and g are arbitrary functions. (Note that $Q(x,t)$ may not be defined for some values of x or t.) Using (1.4), define a generalized approximate parametric derivative of f at x to be

$$f^*(x) = \text{app lim}_{t\to 0} Q(x,t) .$$

The usual combinations of upper, lower, left and right generalized approximate parametric derivatives are defined analogously. For example, the upper right generalized approximate parametric derivate is

$$\overline{f}^{*+}(x) = \text{app lim sup}_{t\to 0+} Q(x,t) .$$

Appropriate choices of a_i, h_i, g and n allow $Q(x,t)$ to be the difference quotient of many different generalized derivatives. Examples of this will be presented in Section 3.

§2. In this section it is assumed that g is a specified function, $h_i,\ldots,h_n$ are fixed parameter functions and $a_1,\ldots,a_n$ are nonzero constants. Given a function f, it is then clear that $Q(x,t)$ is determined by (1.4). The main theorem of this paper is

THEOREM 1. If f is a measurable function such that f^{*+} exists everywhere, finite or infinite, then f^{*+} is in Baire class one.

The proof of this theorem depends upon the following series of lemmas.

LEMMA 1. Let M be a bounded measurable set with $g \neq 0$ on M. If f is a measurable function and h is a parameter function, then given $\epsilon > 0$ and $\delta > 0$ there exists $r > 0$ such that $0 < |x| < r$ implies

$$|\{t \in M : |\frac{f(x+h(t))-f(h(t))}{g(t)}| < \delta\}| > |M| - \epsilon .$$

Proof. It follows from Lusin's theorem and (1.2) that there exists a compact set $P \subset h(M)$ such that $f|_P$ is continuous and $|h^{-1}(P)| > |M| - \epsilon$. According to (1.1) and Lusin's theorem there exists a closed set $Q \subset h^{-1}(P)$ such that $g|_Q$ is continuous and $|Q| > |M| - \epsilon$. Since $g \neq 0$ on Q and Q is compact, there exists $m > 0$ such that $m = \min\{|g(x)| : x \in Q\}$.

The absolute continuity of the Lebesgue integral guarantees that $\lim_{r\to 0}|(r+P) \cap P| = |P|$. This and (1.3) imply that $\lim_{r\to 0}|h^{-1}((r+P)\Delta P)| = 0$. Thus, there exists $r > 0$ such that $0 < |x| < r$ implies

$$|h^{-1}((x+P) \cap P) \cap Q| > |M| - \epsilon .$$

In addition, since $f|_P$ is continuous and P is compact, r may be chosen small enough so that if $x,y \in P$ with $|x-y| < r$, then $|f(x)-f(y)| < m\delta$.

Finally, let $0 < |x| < r$. Then

$$|\{t \in M : |\frac{f(x+h(t))-f(h(t))}{g(t)}| < \delta\}| \geq |\{t \in Q : |\frac{f(x+h(t))-f(h(t))}{g(t)}| < \delta\}|$$

$$\geq |\{t \in Q : h(t) \in P \text{ and } x+h(t) \in P\}| \geq |h^{-1}((x+P) \cap P) \cap Q| > |M| - \epsilon .$$

LEMMA 2. Let $a > 0$, $b \in (0,1)$ and $c \in \mathbb{R}$. Then

$$G(a,b,c) = \{x : |\{t \in (0,a) : Q(x,t) > c\}| > ab\}$$

is open.

Proof. Without loss of generality it may be assumed that $0 \in G(a,b,c)$. Thus, if $S = \{t \in (0,a) : Q(0,t) > c\}$, then $|S| > ab$. There exists $\delta > 0$ small enough so that if $M = \{t \in (0,a) : Q(0,t) > n\delta + c\}$, then $|M| > ab$. Let $\epsilon > 0$ be such that $n\epsilon = |M| - ab$. Apply Lemma 1, with M, ϵ, and δ as in Lemma 1 and $a_i f$ taking the place of f to get an $r_i > 0$ such that $0 \leq |x| < r_i$ implies

$$|\{t \in M : \left|\frac{a_i f(x+h_i(t))-f(h_i(t))}{g(t)}\right| < \delta\}| > |M| - \epsilon .$$

Let $r = \min\{r_1,\dots,r_n\}$. If $0 \leq |x| < r$, then

$$|\{t\in(0,a): Q(x,t)>c\}| \geq |\{t\in M: |Q(x,t)-Q(0,t)|<n\delta\}|$$

$$\geq |\{t\in M: n\delta > \Sigma_{i=1}^{n}\left|\frac{a_i f(x+h_i(t))-f(h_i(t))}{g(t)}\right|\}|$$

$$\geq |\cap_{i=1}^{n}\{t\in M: \left|\frac{a_i f(x+h_i(t))-a_i f(h(t))}{g(t)}\right| < \delta\}| > |M|-n\epsilon > ab\ .$$

From this we conclude that $x\in G$ and therefore $(-r,r)\subset G$. This implies that G is open.

Of course, by considering $-f$ in place of f in the proof of Lemma 2, a companion lemma arises.

LEMMA 2'. Let $a>0$, $b\in(0,1)$ and $c\in\mathbb{R}$. Then

$$H(a,b,c) = \{x: |\{t\in(0,a): Q(x,t)<c\}| > ab\}$$

is open.

Now, suppose f is a function such that f^{*+} exists everywhere. In order to show that f^{*+} is in Baire class one, it suffices to show that

$$\{x: f^{*+}(x)\geq c\} \text{ and } \{x: f^{*+}(x)\leq c\}$$

are both G_δ sets for every $c\in\mathbb{R}$. To see this, suppose $\frac{1}{2}<b<1$ and $m\in\mathbb{N}$. Then if

$$x\in\cap_{k=1}^{\infty}\cup_{n=k}^{\infty} G(\tfrac{1}{n},b,c-\tfrac{1}{m})$$

there exist a sequence of integers, $n_1,n_2,\ldots,$ such that

$$x\in G(\frac{1}{n_i},b,c-\frac{1}{m})$$

for each i. This implies

$$|\{t\in(0,\frac{1}{n_i}): Q(x,t)>c-\frac{1}{m}\}| > \frac{b}{n_i}\ ,$$

which, in turn, implies that $f^{*+}(x)>c-\frac{1}{m}$. Therefore,

$$\cap_{k=1}^{\infty}\cup_{n=k}^{\infty} G(\tfrac{1}{n},b,c-\tfrac{1}{m}) \subset \{x: f^{*+}(x)>c-\tfrac{1}{m}\}\ .$$

Intersecting over all values of $m\in\mathbb{N}$, it follows that

$$(2.1)\qquad \cap_{m=1}^{\infty}\cap_{k=1}^{\infty}\cup_{n=k}^{\infty} G(\tfrac{1}{n},b,c-\tfrac{1}{m}) \subset \{x: f^{*+}(x)\geq c\}\ .$$

On the other hand, if $f^{*+}(x)\geq c$, then for each $m\in\mathbb{N}$ there is an N large enough so that when $n\geq N$

$$|\{t\in(0,\tfrac{1}{n}): Q(x,t)>c-\tfrac{1}{m}\}| > \tfrac{b}{n}\ .$$

From this it follows that $x\in\cup_{n=k}^{\infty}G(\frac{1}{n},b,c-\frac{1}{m})$ for all k. This, in turn, implies $x\in\cap_{k=1}^{\infty}\cup_{n=k}^{\infty}G(\frac{1}{n},b,c-\frac{1}{m})$. Since $f^{*+}(x)\geq c$, this same reasoning must hold for any $m\in\mathbb{N}$. Therefore, $f^{*+}(x)\geq c$ implies

(2.2) $$x \in \bigcap_{m=1}^{\infty} \bigcap_{k=1}^{\infty} \bigcup_{n=k}^{\infty} G(\tfrac{1}{n}, b, c - \tfrac{1}{m}).$$

From (2.1) and (2.2) it follows that

(2.3) $$\{x : f^{*+}(x) \geq c\} = \bigcap_{m=1}^{\infty} \bigcap_{k=1}^{\infty} \bigcup_{n=k}^{\infty} G(\tfrac{1}{n}, b, c - \tfrac{1}{m}).$$

An application of Lemma 2 now shows that $\{x : f^{*+}(x) \geq c\}$ is a G_δ set for every $c \in \mathbb{R}$. A similar argument, using Lemma 2' yields that $\{x : f^{*+}(x) \leq c\}$ is a G_δ set for every $c \in \mathbb{R}$. Now the standard theorem in Goffman [G, p.141] implies f^{*+} is in Baire class one and Theorem 1 is proved.

An argument similar to the one given above is contained in Larson [L1], where it is shown that the first approximate symmetric derivative of a measurable function is in Baire class one.

The following two corollaries are immediately seen to be true from Theorem 1.

COROLLARY 1. If f is a function such that f^* exists everywhere, then f^* is in Baire class one.

COROLLARY 2. If f is a function such that f^{*-} exists everywhere, then f^{*-} is in Baire class one.

In the case of the extreme approximate parametric derivates, the following theorem can be proved.

THEOREM 2. If f is a function and $c \in [-\infty, \infty]$, then

$$\{x : \overline{f}^{*+}(x) > c\},\ \{x : \overline{f}^{*-}(x) > c\},\ \{x : \underline{f}^{*+}(x) < c\} \text{ and } \{x : \underline{f}^{*-}(x) < c\}$$

are all $G_{\delta\sigma}$ sets and

$$\{x : \overline{f}^{*+}(x) < c\},\ \{x : \overline{f}^{*-}(x) < c\},\ \{x : \underline{f}^{*+}(x) > c\} \text{ and } \{x : \underline{f}^{*-}(x) > c\}$$

are all $F_{\sigma\delta\sigma}$ sets.

Proof. Choose any $b \in (0,1)$ and any $c \in \mathbb{R}$. If

$$x \in \bigcap_{k=1}^{\infty} \bigcup_{0 < a < 1/k} G(a,b,c),$$

then for each $k \in \mathbb{N}$ there is an $a_k \in (0, \frac{1}{k})$ such that

$$|\{t \in (0,a_k) : Q(x,t) > c\}| > a_k b.$$

This easily implies that $\overline{f}^{*+}(x) > c$. From this, it follows that if

$$x \in \bigcup_{n=2}^{\infty} \bigcap_{k=1}^{\infty} \bigcup_{0 < a < 1/k} G(a, 1/n, c),$$

then $\overline{f}^{*+}(x) > c$. Therefore

(2.4) $$\bigcup_{n=2}^{\infty} \bigcap_{k=1}^{\infty} \bigcup_{0 < a < 1/k} G(a, 1/n, c) \subset \{x : \overline{f}^{*+}(x) > c\}.$$

If $\overline{f}^{*+}(x) > c$, then there is an $n_0 \in \mathbb{N}$ large enough so that

$$\text{app}\limsup_{a\to 0} \frac{|\{t \in (0,a) : Q(x,t) > c\}|}{a} > \frac{1}{n_0}.$$

From this it can be seen that for each $k \in \mathbb{N}$ there exists $a_k \in (0,1/k)$ such that $x \in G(a_k, 1/n_0, c)$. Therefore

$$x \in \bigcap_{k=1}^{\infty} \bigcup_{0<a<1/k} G(a,1/n_0,c) \subset \bigcup_{n=2}^{\infty} \bigcap_{k=1}^{\infty} \bigcup_{0<a<1/k} G(a,1/n,c).$$

This yields

$$\{x : \overline{f}^{*+}(x) > c\} \subset \bigcup_{n=2}^{\infty} \bigcap_{k=1}^{\infty} \bigcup_{0<a<1/k} G(a,1/n,c). \tag{2.5}$$

Comparing (2.4) and (2.5) it is seen that

$$\{x : \overline{f}^{*+}(x) > c\} = \bigcup_{n=2}^{\infty} \bigcap_{k=1}^{\infty} \bigcup_{0<a<1/k} G(a,1/n,c).$$

Lemma 2 now implies that $\{x : \overline{f}^{*+}(x) > c\}$ is a $G_{\delta\sigma}$ set.

If $|c| = \infty$, then $\{x : \overline{f}^{*+}(x) > \infty\} = \emptyset$ and $\{x : \overline{f}^{*+}(x) > -\infty\} = \bigcup_{n=1}^{\infty} \{x : \overline{f}^{*-}(x) > -n\}$ are both $G_{\delta\sigma}$ sets. By considering $f(-x)$, $-f(x)$ and $-f(-x)$ in the above argument, it can be seen that $\{x : \overline{f}^{*-}(x) > c\}$, $\{x : \underline{f}^{*+}(x) < c\}$ and $\{x : \underline{f}^{*-}(x) < c\}$ are $G_{\delta\sigma}$ sets, respectively.

The second part of the theorem follows from the observations that

$$\{x : \overline{f}^{*+}(x) \geq c\} = \bigcap_{n=1}^{\infty} \{x : \overline{f}^{*+}(x) > c - \tfrac{1}{n}\}$$

and

$$\{x : \overline{f}^{*+}(x) = \infty\} = \bigcap_{n=1}^{\infty} \{x : \overline{f}^{*+}(x) > n\},$$

which are both $G_{\delta\sigma\delta}$ sets. Theire complements are the sets of the theorem.

Using Goffman [G, p.141] again, the following corollary is immediate.

COROLLARY 3. If f is a measurable function, then its extreme approximate parametric derivates are in Baire class three.

3. This section contains some consequences of Theorems 1 and 2. The examples presented here by no means exhaust the possibilities, but rather, are meant to point out some novel or interesting results.

a) Approximate continuity. The simplest nontrivial generalized difference quotient which satisfies the conditions of Theorems 1 and 2 is $Q_0(x,t) = f(x+t)$. As a first, easy example, we state the following theorem, which certainly must be known, although we cannot find a reference for it.

THEOREM 3. Let f be an arbitrary function which is right approximately continuous at every point. Then f is in Baire class one.

Proof. The fact that f is right approximately continuous at each point is equivalent to the statement that $f(x) = \text{app}\lim_{h\to 0+} Q_0(x,h)$. Consequently, the theorem follows from Theorem 1, if it can be shown that f is measurable. To do this, let $a \in \mathbb{R}$ and $E = \{x : f(x) > a\}$. It is clear that E has lower right density equal to one at each of its points. Consideration of the characteristic function of E and the approximate version of the Denjoy-Young-Saks Theorem (S. Chow [C].) shows the lower left density of E is one a.e. on E. Thus, E has density one a.e. and must be measurable. A proof that $\{x : f(x) < a\}$ is measurable runs along similar lines. From these observations, it is easy to see that f is measurable.

It is well-known that an approximately continuous function is in Baire class one; see, for example, Zahorski [Z].

A function is called approximately upper semicontinuous iff for every $x \in \mathbb{R}$, $f(x) = \text{app}\limsup_{h\to 0} Q_0(x,h)$. The following theorem is immediate from Theorem 2.

THEOREM 4. Let f is a measurable function which is approximately upper semicontinuous. Then f is in Baire class three.

b) Symmetric differentiability. Perhaps the simplest generalized difference quotient which satisfies (1.4) and is generally recognized as a difference quotient is that of the symmetric derivative, $[f(x+h)-f(x-h)]/2h$. One of the standard ways to extend the first symmetric derivative to higher orders is to use the Riemann derivative, which has difference quotient

$$S_n(f,x,h) = (2h)^{-n} \Sigma_{k=0}^{n} (-1)^k \binom{n}{k} f(x+(n-2k)h) . \tag{3.1}$$

(For the other common method, due to de la Vallée Poisson, see Zygmund [Z3, p.59].) The n'th approximate Riemann derivative of f is defined as

$$f_{(n)}^{ap}(x) = \text{app}\lim_{h\to 0} S_n(f,x,h) .$$

The corresponding upper and lower derivates are denoted $\overline{f}_{(n)}^{ap}$ and $\underline{f}_{(n)}^{ap}$. The normal Riemann derivatives and derivates are written $f_{(n)}$, $\underline{f}_{(n)}$, etc. The following theorem follows at once from the definitions and Theorems 1 and 2.

THEOREM 5. If f is any measurable function and n is an odd integer, then $\overline{f}_{(n)}^{ap}$ and $\underline{f}_{(n)}^{ap}$ are in Baire class three. Furthermore, if $f_{(n)}^{ap}$ exists everywhere, finite or infinite, then $f_{(n)}^{ap}$ is in Baire class one.

The case when n=1 has already been noted by this author in Larson [L1]. When n is an even integer, Theorems 1 and 2 do not directly imply anything about $f_{(n)}^{ap}$ because when $k = n/2$ in (3.1) the parameter function vanishes, which violates (1.3). However, the following interesting theorem is true.

THEOREM 6. If n is an even integer and f is a measurable function such that $f_{(n)}^{ap}$ exists and is finite everywhere, then f is in Baire class one.

Proof. Since $f_{(n)}^{ap}$ is finite everywhere, it must be true that for each x,

$$\text{app}\lim_{h\to 0} (2h)^n S_n(f,x,h) = 0 . \tag{3.2}$$

Using (3.1), $S_n(f,x,h)$ may be solved for f(x), and then (3.2) implies

$$f(x) = \text{app}\lim_{h\to 0} (-1)^{n/2} \binom{n}{n/2} \Sigma_{\substack{0 \le k \le n \\ k \ne n/2}} (-1)^{k+1} \binom{n}{k} f(x+(n-2k)h) .$$

Now, Theorem 1 implies that f is in Baire class one.

A function f is approximately symmetric iff for each $x \in \mathbb{R}$,

$$\text{app}\lim_{h\to 0} h^2 S_2(f,x,h) = 0 .$$

The following corollary is proved similarly to Theorem 6.

COROLLARY 4. If f is a measurable, approximately symmetric function, then f is in Baire class one.

The assumption of measurability is necessary in Corollary 4, for there do exist nonmeasurable symmetric functions. The well-known nonmeasurable linear functions provide examples.

In Weil [W], it is shown that if f is a function of Baire class one with the Darboux property such that $\underline{f}_{(2)}(x) \ge 0$ everywhere, then f is convex. Combining this with Theorem 6, the following corollary arises.

COROLLARY 5. If f is a measurable function with the Darboux property such that $f_{(2)}(x) \ge 0$ everywhere, then f is convex.

c) Ordinary approximate derivatives. Suppose f is an arbitrary function such that the ordinary approximate derivative, f'_{ap}, exists everywhere, finite or infinite. According to Theorem I(iii) in Chow [C], f'_{ap} is finite a.e., and therefore f is approximately continuous a.e. This implies that f is measurable. From Theorem 5, the following theorem is immediate.

THEOREM 7. If f is an arbitrary function such that f'_{ap} exists everywhere, finite or infinite, then f'_{ap} is in Baire class one.

Theorem 7 was first proved by Tolstoff [T] in the case when f'_{ap} is finite valued. Preiss [P] proved it in the more general case given above.

BIBLIOGRAPHY

[C] Shu-Er Chow, "On approximate derivatives.", Bull. Amer. Math. Soc., 54(1948), 793-802.

[E] M. Evans and P. Humke, "Parametric Differentiation," Colloq. Math. 45(1981) #1, 125-131.

[F] F. M. Filipczak, "Sur les derivees symmetriques des fonctions approximativement continues", Coll. Math. 34(1976), 249-256.

[G] C. Goffman, "Real Functions", Rinehart, New York, 1960.

[L1] L. Larson, "The Baire class of approximate symmetric derivates", Proc. Amer. Math. Soc., 87(1), 125-130.

[L2] L. Larson, "The Symmetric Derivative", Trans. Amer. Math. Soc., 277(2), 1983, 589-599.

[N] C. J. Neugebauer, "Smoothness and differentiability in L_p", Stud. Math. 25(1964), 81-91.

[P] D. Preiss, "Approximate derivatives and Baire classes", Czech. Math. J., 21(1971), 373-382.

[T] G. Tolstoff, "Sur la derivee approximative exacte", Rec. Mat. (Mat. Sbornik) N. S. (1938), 499-504.

[W] C. E. Weil, "Monotonicity, convexity and symmetric derivates", Trans. Amer. Math. Soc., 222(1976), 225-237.

[Z1] Z. Zahorski, "Sur la premiere derivee", Trans. Amer. Math. Soc., 69(1950), 1-54.

[Z2] A. Zygmund, "Trigonometric Series", Vol. 1, Cambridge Univ. Press, 1959.

[Z3] A. Zygmund, "Trigonometric Series", Vol. 2, Cambridge Univ. Press, 1959.

DEPARTMENT OF MATHEMATICS
UNIVERSITY OF LOUISVILLE
LOUISVILLE, KY 40292

Contemporary Mathematics
Volume 42, 1985

ON GENERALIZATIONS OF EXACT PEANO DERIVATIVES

Cheng-Ming Lee

ABSTRACT. The exact Peano derivative and the processes by which it is generalized are discussed. A new generalization, the ultimate Peano derivative, is given and open problems are noted.

1. INTRODUCTION. It is a fundamental result in calculus that every finite derivative on an interval determines its primitive up to a term of constant. However, the question of how to reconstruct a primitive from its derivative had remained an open problem for a long time. The first solution to this primitive problem was given by A. Denjoy in 1912 [6]. He introduced a totalization procedure by a transfinite induction, using the Lebesgue integral and the Cauchy and Harnack extension processes. This procedure gives rise to an integral now known as the Denjoy integral in the restricted sense. Since it is equivalent to the Perron integral by the Hake-Alexandroff-Looman theorem, this integral is also called the Denjoy-Perron integral [19].

The impact of solving a long standing open problem is lasting and unmeasurable. Denjoy's solution to the primitive problem is no exception. It has motivated and inspired many authors to publish many interesting works. For example, there are more than 350 publications by about 125 authors quoted in the survey article by P.S. Bullen in [5], where the development of nonabsolute integral theory since Denjoy's solution is concisely and beautifully discussed. Here we will briefly indicate some of the recent results related to exact Peano derivatives and discuss some of the problems which are open for further study.

2. EXACT n^{th} PEANO DERIVATIVES. Recall that a function F is called to have a kth Peano derivative at x if there exist k numbers $\alpha_1, \alpha_2, \ldots, \alpha_k$ such that

$$F(x+h) = F(x) + h\alpha_1 + \frac{h^2}{2!}\alpha_2 + \ldots + \frac{h^k}{k!}\alpha_k + o(h^k)$$

as $h \to 0$, and in this case the number ${}_k$ denoted as $F_{(k)}(x)$ is called the kth Peano derivative of the function F at x. The first Peano derivative $F_{(1)}(x)$ is just the derivative $F'(x)$. If the repeated k^{th} derivative $F^{(k)}(x)$ exists,

1980 Mathematics Subject Classification. 26A24.

then so does $F_{(k)}(x)$ and the two are equal. However, it may happen that $F_{(k)}(x)$ exists while $F^{(k)}(x)$ does not exist for $k > 1$. Thus, the kth Peano derivative is a generalization of the repeated k^{th} derivative for $k > 1$. A finite function f is called an exact k^{th} Peano derivative on an interval if $f = F_{(k)}$ on the interval for some function F. For many interesting studies of such k^{th} Peano derivatives and some of its generalizations we refer to the recent survey article by M. Evans and C. Weil [8].

It can be proved that an exact k^{th} Peano derivative $F_{(k)}$ determines the function F up to a polynomial of degree $k-1$ on the interval concerned. The problem of reconstructing the function F up to such a polynomial from its kth Peano derivative $F_{(k)}$ has been solved by A. Denjoy [7] using a generalized k^{th} totalization process, and by R.D. James [10] and P.S. Bullen [4] and many others using an extended Perron method. For $k > 1$, their processes give rise to or are integrals of order greater than 1, and will not be discussed here. Instead, we will consider the C_{k-1} P-integral in the scale of Cesàro-Perron integrals due to J.C. Burkill in [2], [3], which is an integral of first order and recaptures the $(k-1)^{th}$ Peano derivative $F_{(k-1)}$ (up to a constant) from the kth Peano derivative $F_{(k)}$.

The C_nP-integral in Burkill's scale for $n=0,1,2,\ldots$, is defined by induction. The C_0P-integral is taken to be the Denjoy-Perron integral, and for $n \geq 1$, the C_nP-integral is defined by a Perron method using the concepts of the nth Cesàro continuity and the nth Cesàro derivates of $C_{n-1}P$-integrable functions. For a $C_{n-1}P$-integrable function f on the compact interval $[a,b]$, one shows that the function F defined by $F(x) = (b-x)^{n-1}f(x)$ is also $C_{n-1}P$-integrable on $[a,b]$, and the following number C_n is called the nth Cesàro mean of the function f on $[a,b]$:

$$C_n + C_n(f;a,b) = n(b-a)^n \int_a^b F(x)dx.$$

A function f is said to be C_n-continuous at x if it is $C_{n-1}P$-integrable on a neighborhood of x and

$$\lim_{h\to 0} C_n(f;x,x+h) = f(x).$$

The lower n^{th} Cesàro derivate of a $C_{n-1}P$-integrable function f at the point x is defined as

$$\ell C_nDf(x) = \liminf_{h\to 0}(n+1)h^{-1}[C_n(f;x,x+h)-f(x)].$$

Instead of going through the process to define the C_nP-integral, we only remark that the C_nP-integral can be characterized by directly using the concept of Peano derivates. This seems to be a fact implicitly indicated in Burkill's original work [2], [3], but it was only recently made explicit by J. Marik and J.A. Bergin [1], [12]. We list the essential ingredient for this characterization as a theorem which will be the basis for our later discussion.

THEOREM 1. For a finite function f to be an exact n^{th} Peano derivative on a compact interval I it is necessary and sufficient that f is C_n-continuous at each point of I, i.e. f is $C_{n-1}P$-integrable on $[a,b]$ and

$$\lim_{h\to 0} C_n(f;x,x+h) = f(x) \quad \text{for each } x \text{ in } I .$$

Furthermore, if f is the exact n^{th} Peano derivative of F on the interval I, then

$$\ell F_{(n+1)}(x) = \ell C_n Df(x)$$

for each x in I, where $\ell F_{(n+1)}(x)$ denotes the lower $(n+1)^{th}$ Peano derivate of F at x.

REMARK 1. The C_nP-integral for $n > 1$ defined by means of the Perron method is equivalent to the C_nD-integral of W.L.C. Sargent [20] defined by means of a Denjoy method. Some defects in both [3] and [20] were later repaired by S. Verblunsky in [21], [22].

REMARK 2. The first part of Theorem 1 can be viewed as an extension of Denjoy's solution for the derivatives to the exact n^{th} Peano derivatives. On the other hand, the exact derivatives have been characterized by C.J. Neugebauer [18] in another interesting way. Whether the exact n^{th} Peano derivatives can be characterized in a similar manner is still an open question.

REMARK 3. The proof of Theorem 1 is based on the integration by parts formula for the C_nP-integral for $0 \le k \le n-1$, and the proof of the formula is closely related to the consideration of the Peano derivates of the product of two functions. This, in turn, for $n=1$, is closely related to the multipliers of exact derivatives, which have been characterized recently by R.J. Fleissner [9]. The question of how to extend Fleissner's result for the multipliers of the exact n^{th} Peano derivatives still remains to be answered.

3. EXACT ABSOLUTE PEANO DERIVATIVES. A natural way to extend the C_n-continuous functions on an interval is to let the n to vary with the point in the interval. Thus, we say that a function f is pointwise Cesàro continuous on an interval if for each x in the interval there exists a positive integer n such that f is C_n-continuous at x. This notion happens to be closely related to the exact absolute Peano derivatives studied earlier by M. Laczkovich in [11]. A function F is said to have an absolute Peano derivative at x if there exist a non-negative integer k and a function G such that $G_{(k)} = F$ on a neighborhood of x and $G_{(k+1)}(x)$ exists (and is finite), and in this case, the number $G_{(k+1)}(x)$ is unambiguously denoted as $F^{(*)}(x)$ and is called the absolute Peano derivative of F at x. A function f is an exact absolute Peano derivative on an interval if there exists a function F such that

$F^{(*)} = f$ on the interval. Many interesting properties for the exact n^{th} Peano derivatives listed in [8] hold good for the exact absolute Peano derivatives [11], [15].

THEOREM 2. If f is pointwisely Cesàro continuous on a compact interval, then f is an exact absolute Peano derivative on the interval.

REMARK 4. We suspect that the converse of Theorem 2 is true. One of the questions related to this converse problem is the following: Are there any exact absolute Peano derivatives on a compact interval which are not C_nP-integrable for every positive integer n ?

4. EXACT n^{th} GENERALIZED PEANO DERIVATIVES. A function F continuous on a neighborhood of a point x may not have the nth Peano derivative at x, but there may exist a positive integer k such that a kth primitive G of F in that neighborhood has the $(k+n)^{th}$ Peano derivative at x. If such a positive integer k exists, the number $G_{(k+n)}(x)$ is denoted unambiguously as $F_{[n]}(x)$ and is called the nth generalized Peano derivative of F at x. Of course, a function f is said to be an exact n^{th} generalized Peano derivative on an interval if there exists a function F such that $F_{[n]} = f$ on the interval. The following result can be proved easily using the definitions and a standard compactness argument (cf.[14] for an indirect proof).

THEOREM 3. If f is an exact absolute Peano derivative on a compact interval, then there exists a positive integer n such that f is an exact n^{th} generalized Peano derivative on the interval.

REMARK 5. Every exact 1st generalized Peano derivative is an absolute Peano derivative. But for $n \geq 2$, we conjecture that it is possible to have an exact n^{th} generalized Peano derivative which is not an exact absolute Peano derivative.

REMARK 6. If in the definition of the absolute Peano derivative, the Peano derivative $G_{(k)}$ and $G_{(k+1)}(x)$ are replaced by the generalized $G_{[k]}$ and $G_{[k+1]}(x)$, one obtains the notion called the absolute generalized Peano derivative. However, on a compact interval every exact absolute generalized Peano derivative is an exact n^{th} generalized Peano derivative [14] and hence nothing new is obtained through this process. In the next section we will describe another process, which may produce something new.

REMARK 7. Pointwisely Cesàro continuous functions and exact n^{th} generalized Peano derivatives were initiated in [13], where we were concerned with the generalization of the Cesàro-Perron integrals, i.e. the C_nP-integrals for $n=0,1,2,\ldots$. An integrable function f can be thought as linear functional on a certain function space which contains the kernel $(b-x)^{n-1}$ used in the definition of the C_n-mean for $n=1,2,3,\ldots$. Thus, we are led to consider the

"C_∞-continuous" functions as some particular linear functionals. It happens that the concept of values of a Schwartz distribution at points as studied by S. Lojasiewicz [16] can be used to specify this particular linear functionals. More precisely, we have the following result which is essentially due to Lojasiewicz [16] (cf. [13]).

THEOREM 4. For a finite function f to be an exact generalized Peano derivative on an interval I it is necessary and sufficient that there exists a distribution T of finite order on the interval I such that $v(T;x) = f(x)$ for all x in I, where $v(T;x)$ denotes the value of the distribution T at the point x.

Here, a distribution T on a linear interval is said to have a value at the point x_0 of the interval if the distribution $T(ax+x_0)$ converges in the distributional sense as $a \to 0$, i.e. for each test function ϕ on the interval,

$$\langle T, \frac{1}{a}\,\phi(\frac{x-x_0}{a})\rangle$$

converges as $a \to 0$. Note that if $T(ax+x_0)$ converges in the distributional sense as $a \to 0$, then the limit is a constant distribution in a neighborhood of x_0, and this constant is denoted as $v(T;x_0)$ and is called the value of the distribution T at the point x_0.

5. ULTIMATE PEANO DERIVATIVES. Let f be a function continuous on a neighborhood of x such that the $(n-1)^{th}$ generalized Peano derivative $f_{[n-1]}(x)$ exists (and is finite). For each non-negative integer k and for each $h \neq 0$, let

$$\varepsilon^k_{n-1}(f;x,x+h) = [G(x+h)-G(x)-\sum_{i=1}^{k+n-1} h^i G_{[i]}(x)/i!](k+n-1)1/h^{k+n-1},$$

where G is a kth primitive of f on the neighborhood of x. Note that ε^k_{n-1} as defined is independent of which k^{th} primitive is taken as G. Now define

$$\ell_k f_{[n]}(x) = \liminf_{h\to 0} \varepsilon^k_{n-1}(f;x,x+h)(k+n)/h ,$$

and

$$u_n f_{[n]}(x) = \limsup_{h\to 0} \varepsilon^k_{n-1}(f;x,x+h)(k+n)/h .$$

Then one shows that

$$\ell_k f_{[n]}(x) \le \ell_{k+1} f_{[n]}(x) \le u_{k+1} f_{[n]}(x) \le u_k f_{[n]}(x) .$$

Hence $\lim_{k\to\infty} \ell_k f_{[n]}(x)$ and $\lim_{k\to\infty} u_k f_{[n]}(x)$ exist and will be denoted as $\ell_\infty f_{[n]}(x)$ and $u_\infty f_{[n]}(x)$, respectively. It is easily proved that the nth generalized Peano derivative exists if and only if there exists a positive integer k such that

$$\ell_k f_{[n]}(x) = u_k f_{[n]}(x) \quad (\neq \pm \infty)$$

and in this case one has

$$f_{[n]}(x) = \ell_j f_{[n]}(x) = u_j f_{[n]}(x) \quad \text{for } k \leq j \leq \infty.$$

In particular, if $f_{[n]}(x)$ exists, then

$$\ell_\infty f_{[n]}(x) = u_\infty f_{[n]}(x).$$

However, we conjecture that the converse may fail to hold. If that is so and the last equality holds, the common value, denoted by $_\infty f_{[n]}(x)$, will be called the nth ultimate Peano derivative of f at x, as suggested by C. Weil. Whether the conjecture is correct remains to be seen. If it is correct, properties and characterizations of such ultimate Peano derivatives will be topics for further study.

BIBLIOGRAPHY

1. J.A. Bergin, "A new characterization of Cesàro-Perron integrals using Peano derivatives", Trans. Amer. Math. Soc., 228 (1977), 287-305.

2. J.C. Burkill, "The Cesàro-Perron integral", Proc. London Math. Soc., (2) 34 (1932), 314-322.

3. ______, "The Cesàro-Perron scale of integration", Proc. London Math. Soc., (2) 39 (1935), 541-552.

4. P.S. Bullen, "The P^n-integral", J. Austral. Math. Soc., 14 (1972), 219-236.

5. ______, "Non-absolute integrals: a survey", Real Analysis Exchange, 5 (1979-80), 195-259.

6. A. Denjoy, "Une extension de l'integrale de M. Lebesgue", C. R. Acad. Sci. Paris, 154 (1912), 1075-1078.

7. ______, "Sur l'intégration des coefficients différentials d'ordre supérieur", Fund. Math., 25 (1935), 273-326.

8. M.J. Evans and C.E. Weil, "Peano derivatives: a survey", Real Analysis Exchange, 7 (1982-83), 5-23.

9. R.J. Fleissner, "Multiplication and the fundamental theorem of calculus: a survey", Real Analysis Exchange, 2 (1976), 7-34.

10. R.D. James, "Generalized n^{th} primitives", Trans. Amer. Math. Soc., 76 (1954), 149-176.

11. M. Laczkovich, "On the absolute Peano derivatives", Ann. Univ. Sci. Budapest, Eötvös Sect. Math., 21 (1978), 83-97.

12. C.-M. Lee, "An approximate extension of Cesàro-Perron integrals", Bull. Inst. Acad. Sinica, 4 (1976), 73-82.

13. ______, "Generalizations of Cesàro continuous functions and integrals of Perron type", Trans. Amer. Math. Soc., 266 (1981), 461-481.

14. ______, "On generalized Peano derivatives", Trans. Amer. Math. Soc., 275 (1983), 381-396.

15. C.-M. Lee, "On absolute Peano derivatives", Real Analysis Exchange, 8 (1982-83), 227-243.

16. S. Lojasiewicz, "Sur la valeur et la limite d'une distribution en un point", Studia Math., 16 (1957), 1-36.

17. J. Mikusinski and R. Sikorski, "The Elementary Theory of Distributions, I", Rozprawy Mat., 12 (1957), 54 pp.

18. C.J. Neugebauer, "Darboux functions on Baire class one and derivatives", Proc. Amer. Math. Soc., 13 (1962), 838-843.

19. S. Saks, "Theory of Integral", Monogr. Mat., 7, Warszawa-Lwów, 1937.

20. W.L.C. Sargent, "A descriptive definition of Cesàro-Perron integrals", Proc. London Math. Soc., (2) 52 (1941), 212-247.

21. S. Verblunsky, "On the Peano derivatives", Proc. London Math. Soc., (3) 22 (1971), 313-324).

22. ______, "On a descriptive definition of Cesàro-Perron integrals", J. London Math. Soc., (2) 3 (1971), 326-333.

DEPARTMENT OF MATHEMATICAL SCIENCES
UNIVERSITY OF WISCONSIN-MILWAUKEE
MILWAUKEE, WISCONSIN 53201

Contemporary Mathematics
Volume 42, 1985

BEST MONOTONE APPROXIMATION IN $L_\infty[0,1]$

David Legg

ABSTRACT. The Polya algorithm is shown to converge for a function class larger than the quasicontinuous functions. Further examples of functions for which convergence facts are given.

1. INTRODUCTION. If f(x) is Lebesgue measurable on [a,b], let $f_p(x)$ denote the best L_p-approximant to f(x) by nondecreasing functions. This means

$$\|f-f_p\|_p = \inf_g\{\|f-g\|_p\}$$

where g(x) is a nondecreasing function on [a,b]. It is well known that $f_p(x)$ is uniquely determined (up to a.e. equivalence) if $1 < p < \infty$. In [4], Darst and Sahab proved that if f(x) is quasicontinuous (definition to follow), then

$$\lim_{p\to\infty} f_p(x) \equiv f_\infty(x)$$

exists uniformly on [a,b].

A function is quasicontinuous if

$$\lim_{y\to x^+} f(y) \text{ exists at each } x \in [a,b)$$

and

$$\lim_{y\to x^-} f(y) \text{ exists at each } x \in (a,b] .$$

Hence we are provided with a canonical "best" best L_∞-approximant to f(x) by nondecreasing functions if f(x) is quasicontinuous. This procedure for constructing "best" best L_∞ approximants is known as the Polya algorithm and is known to converge in a number of different settings. It is also known that this algorithm will fail to converge in certain settings. See the references [1], [2], [6], [7], and [8] for further details.

1980 Mathematics Subject Classification. 41A30.

In the present paper, we show that the result of Darst and Sahab mentioned above can be extended to include "reasonable" Lebesgue measurable functions of a more general type than quasi-continuous. At the same time, the proof is such that the limit function $f_\infty(x)$ is easier to visualize even when $f(x)$ is continuous.

An example of a Lebesgue measurable function defined on [0,2] such that the Polya algorithm fails to converge is given in [3]. Since then, an example of a function continuous on [0,1), continuous on [1,2], and approximately continuous at 1 for which the Polya algorithm fails to converge has been found. See the paper by Darst in this volume. In this paper, we will present yet another counterexample to the convergence of the Polya algorithm. Let $\mathcal{M}$ be a σ-lattice of Borel subsets of [a,b], and let $f_p(x)$ be the best L_p-approximant to $f(x)$ by $\mathcal{M}$-measurable functions. We show that $\lim_{p\to\infty} f_p(x)$ may fail to exist a.e. even for continuous $f(x)$.

In section 2 we present the construction of $f_\infty(x)$ but omit the proof that $\lim_{p\to\infty} f_p(x) = f_\infty(x)$ since it appears in [3]. In section 3 we present the counterexample in detail since it appears here for the first time. The author wishes to acknowledge D. W. Townsend and R. B. Darst for their contributions to this paper.

2. THE CONSTRUCTION OF $f_\infty(x)$. Let

$$f(x) = \sum_{i=1}^{N} a_i \chi_{E_i}(x)$$

be a Lebesgue measurable simple function. For convenience we assume $a_i - a_j \neq a_k - a_m$ for all $(i,j) \neq (k,m)$. We construct partition points of [0,1] according to the following steps.

Step 1. Let

$$b_{11} = \operatorname*{ess\,max}_{x<y} (f(x)-f(y))^+ .$$

If $b_{11} = 0$, then f is essentially nondecreasing on [0,1] with essential jump discontinuities at $\{z_{111},\ldots,z_{11k}\}$.

If $b_{11} > 0$, then let

$$x_{11} = \inf\{x\colon \exists y_0 > x \ni f(x)-f(y_0) = b_{11} \text{ and } m(t<x \mid f(t)-f(y_0) = b_{11}) > 0\}$$

and

$$y_{11} = \sup\{y > x_{11}\colon \exists x_0 < y \ni f(x_0)-f(y) = b_{11} \text{ and } m(t>y \mid f(x_0)-f(t) = b_{11}) > 0\} .$$

In the preceding definitions, $m(S)$ denotes the Lebesgue measure of S.

Step 2.1. If $x_{11} = 0$, go to step 2.2. If $x_{11} > 0$, let

$$b_{21} = \operatorname*{ess\,max}_{x < y < x_{11}} (f(x) - f(y))^+ .$$

If $b_{21} = 0$, then f is essentially nondecreasing on $[0, x_{11}]$, with essential jump discontinuities at $\{z_{211}, \ldots, z_{21k}\}$.

If $b_{21} > 0$, then let

$$x_{21} = \inf\{x < x_{11} : \exists y_0 > x \ni f(x) - f(y_0) = b_{21} \text{ and } m(t < x \mid f(t) - f(y_0) = b_{21} > 0\}$$

and

$$y_{21} = \sup\{y < x_{11} : \exists x_0 < x_{11},\ x_0 < y \ni f(x_0) - f(y) = b_{21} \text{ and } m(y < t < x_{11} \mid f(x_0) - f(t) = b_{21}) > 0\} .$$

Step 2.2. If $x_{11} = 0$ and $y_{11} = 1$, stop.

If $x_{11} > 0$ and $y_{11} = 1$, go to the next step.

If $x_{11} > 0$ and $y_{11} < 1$, let

$$b_{22} = \operatorname*{ess\,max}_{y_{11} < x < y} (f(x) - f(y))^+ .$$

If $b_{22} = 0$, then f is essentially nondecreasing on $[y_{11}, 1]$, with essential jump discontinuities at $\{z_{221}, \ldots, z_{22k}\}$.

If $b_{22} = 0$ and $b_{21} = 0$, stop.

If $b_{22} = 0$ and $b_{21} > 0$, go to the next step.

If $b_{22} > 0$, then let

$$x_{22} = \inf\{x > y_{11} : \exists y_0 > x \ni f(x) - f(y_0) = b_{22} \text{ and } m(y_{11} < t < x \mid f(t) - f(y_0) = b_{22}) > 0\}$$

and

$$y_{22} = \sup\{y > x_{22} : \exists x_0 > y_{11},\ x_0 < y \ni f(x) - f(y) = b_{22} \text{ and } m(t > y \mid f(x_0) - f(t) = b_{22}) > 0\} .$$

Continue to define the x_{ij}, y_{ij}, and z_{ijk} in this manner over the remaining intervals $[0, x_{21}]$, $[y_{21}, x_{11}]$, $[y_{11}, x_{22}]$ and $[y_{22}, 1]$. Since f is a simple function, this process terminates after finitely many steps.

Let $P = \{x_{ij}\} \cup \{y_{ij}\} \cup \{z_{ijk}\} \cup \{0, 1\}$ and then let $\{t_1, \ldots, t_n\}$ be a relabeling of P in increasing order.

We now define $f_\infty(x)$, which is a best L_∞ approximation to $f(x)$ by nondecreasing functions.

Step 1. If $b_{11}=0$, then f is essentially nondecreasing on $[0,1]$. By the definition of $\{z_{11k}\}$, if $t_i, t_{i+1} \in P$, then f is essentially constant on (t_i, t_{i+1}). Let B_{11i} be that constant.

Then for all $x \in (t_i, t_{i+1})$, define $f_\infty(x) = B_{11i}$ and we are finished.

If $b_{11} > 0$, then $\exists x_{11}^1 > x_{11}$ and $y_{11}^1 < y_{11}$ such that $f(x_{11}^1) - f(y_{11}^1) = b_{11}$. Then for $x \in (x_{11}, y_{11})$ define

$$f_\infty(x) = \frac{1}{2}(f(x_{11}^1) + f(y_{11}^1)) \equiv A_{11}.$$

Step 2.1. If $b_{21}=0$, then f is essentially nondecreasing on $(0, x_{11})$. By the definition of $\{z_{21k}\}$, if $(t_i, t_{i+1}) \subseteq (0, x_{11})$ then f is essentially constant on (t_i, t_{i+1}). Let that constant be B_{21i}. For $x \in (t_i, t_{i+1})$, define

$$f_\infty(x) = \min\{B_{21i}, A_{11}\}.$$

If $b_{21} > 0$, then $\exists x_{21}^1, y_{21}^1$ such that $x_{21} \le x_{21}^1 < y_{21}^1 \le y_{21}$ and $f(x_{21}^1) - f(y_{21}^1) = b_{21}$. Then for all $x \in (x_{21}, y_{21})$, define

$$f_\infty(x) = \min\{\frac{1}{2}(f(x_{21}^1) + f(y_{21}^1)), A_{11}\} \equiv A_{21}.$$

Note that if $A_{21} = A_{11}$, this forces $f_\infty(x) = A_{11}$ for all $x \in (y_{21}, x_{11})$.

Step 2.2. If $b_{22}=0$, then f is essentially nondecreasing on $[y_{11}, 1]$. By the definition of $\{z_{22k}\}$, if $(t_i, t_{i+1}] \subseteq (y_{11}, 1]$, then f is essentially constant on $(t_i, t_{i+1}]$. Let that constant be B_{22i}, and for $x \in (t_i, t_{i+1}]$ define

$$f_\infty(x) = \max\{B_{22i}, A_{11}\}.$$

If $b_{22} > 0$ (and the interval $(x_{22}, y_{22}]$ is defined), then $\exists$ x_{22}^1, y_{22}^1 such that $x_{22} \le x_{22}^1 < y_{22}^1 \le y_{22}$ and $f(x_{22}^1) - f(y_{22}^1) = b_{22}$. Then for all $x \in [x_{22}, y_{22}]$ define

$$f_\infty(x) = \max\{\frac{1}{2}[f(x_{22}^1) + f(y_{22}^1)], A_{11}\} \equiv A_{22}.$$

Step 3.1. If b_{31} is defined and $b_{31}=0$, then f is essentially nondecreasing on $[0, x_3]$ and $f_\infty(x)$ is defined as in step 2.1.

If $b_{31} > 0$, then x_{31} and y_{31} are defined, and $f_\infty(x)$ is defined on $(x_{31}, y_{31}]$ as in step 2.1.

Step 3.2. If b_{32} is defined and $b_{32}=0$, then f is essentially nondecreasing on $[y_{21}, x_{11}]$. By the definition of $\{z_{32}k\}$, if $(t_i, t_{i+1}] \subseteq [x_{21}, x_{11}]$ then f is essentially constant on $(t_i, t_{i+1}]$. Call that constant B_{32i} and for all $x \in (t_i, t_{i+1}]$

define

$$f_\infty(x) = \min[\max\{A_{21}, B_{32i}\}, A_{11}].$$

If $b_{32} > 0$, then $\exists x_{32}^1, y_{32}^1$ such that $x_{32} \le x_{32}^1 < y_{32}^1 \le y_{32}$ and $f(x_{32}^1) - f(y_{32}^1) = b_{32}$.

Then for all $x \in (x_{32}, y_{32}]$ define

$$f_\infty(x) = \min[\max\{A_{21}, \tfrac{1}{2}[f(x_{32}^1) + f(y_{32}^1)]\}, A_{11}].$$

The definition of $f_\infty(x)$ for all subsequent steps follows the patterns established above.

THEOREM 2.1. If $f(x)$ is a simple Lebesgue measurable function and if $\lim_{x \to x_{ij}^+} f(x)$ and $\lim_{x \to y_{ij}^-} f(x)$ exist for each x_{ij}, y_{ij} as described in section 2, then $f_p(x)$ can be chosen so that

$$\lim_{p \to \infty} f_p(x) = f_\infty(x)$$

uniformly on $[0,1]$.

THEOREM 2.2. If f is the uniform limit of a sequence $\{f_n\}$ of simple functions satisfying the hypothesis of Theorem 2.1, then $\lim_{p \to \infty} f_p(x)$ exists uniformly on $[0,1]$.

The proofs of these theorems can be found in [5].

3. A COUNTEREXAMPLE. In this section we construct a continuous function f on $[0,1]$ and a σ-lattice $\mathfrak{M}$ of Borel subsets of $[0,1]$ such that $\lim_{p \to \infty} f_p(x)$ fails to exist on a set of positive measure. Here f_p refers to the best L_p-approximant to f by functions measurable with respect to $\mathfrak{M}$. This generalizes the motion of monotone approximation since nondecreasing functions are precisely those measurable with respect to the particular σ-lattice $\mathfrak{M}_0 = \{(a,1)\} \cup \{[a,1]\}$.

To construct $\mathfrak{M}$ for our counterexample, let $\{A_i\}$ and $\{B_i\}$ be sequences of disjoint closed sets of positive measure and let C be a closed set of positive measure, all in $[0,1]$, with

$$A = \overline{\cup A_i}, \qquad B = \overline{\cup B_i},$$

$$A \cap C = B \cap C = A \cap B = \emptyset,$$

$$\overline{\cup_{i \ge n} A_i} \cap A_j = \emptyset$$

and $$\overline{\cup_{i \ge n} B_i} \cap B_j = \emptyset \quad \text{for } j < n.$$

Let $E_n = (\overline{\cup_{i \ge n} A_i}) \cup (\overline{\cup_{i \ge n} B_i}) \cup C$ and let $\mathfrak{M}$ be the σ-lattice generated by the E_n's and the Borel sets of E_1^c. There exists a continuous function f on $[0,1]$ such that $f|_A = 2$, $f_{|B} = 0$, and $f_{|C} = 1$. Let f_p be the best L_p-approximant to f by functions measurable with respect to $\mathcal{M}$. Note that f_p agrees with f on E_1^c and f_p is increasing with respect to the E_n's.

Now let $a_n = m(\overline{\cup_{i\geq n} A_i})$ and $b_n = m(\overline{\cup_{i\geq n} B_i})$, where m denotes Lebesgue measure.

We will inductively define the a_n's and b_n's such that

$$\frac{a_{2n+1}}{b_{2n+1}} \to \infty$$

and

$$\frac{a_{2n}}{b_{2n}} \to 0$$

fast. In this way, we can make

$$\overline{\lim_{p\to\infty}} f_p \geq 3/2 \text{ on } C$$

and

$$\underline{\lim_{p\to\infty}} f_p \leq 1 \text{ on } C.$$

To begin, let $\epsilon_n \to 0$ and suppose b_1 were to equal 0. Then clearly $\lim_{p\to\infty} f_p = 3/2$ uniformly on E_1. Choose $p_1 > 1$ so that $f_{p_1} > 3/2 - \epsilon_1$ on E_1. Now choose $b_1 > 0$ so small compared to a_1 that we still have $f_{p_1} > 3/2 - \epsilon_1$. In particular, $f_{p_1} > 3/2 - \epsilon_1$ on C.

Now suppose a_2 were to equal 0. Since $b_1 > 0$, we would have $\lim_{p\to\infty} f_p = 1$ uniformly on E_1. Choose $p_2 > 2p_1$ such that $f_{p_2} < 1 + \epsilon_2$ on E_1. Let $0 < b_2 < b_1$ and choose $a_2 > 0$ so small compared to b_2 that we still have $f_{p_2} < 1 + \epsilon_2$ on E_1. In particular, $f_{p_2} < 1 + \epsilon_2$ on C.

Now choose $0 < a_3 < a_2$. If b_3 were to equal 0, we would have $\lim_{p\to\infty} f_p = 3/2$ uniformly on E_3. Let $p_3 > 2p_2$ such that $f_{p_3} > 3/2 - \epsilon_3$ on E_3. Now choose $b_3 > 0$ so small compared to a_3 that we still have $f_{p_3} > 3/2 - \epsilon_3$ on E_3. In particular, $f_{p_3} > 3/2 - \epsilon_3$ on C.

Now suppose a_4 were to equal 0. Since $b_3 > 0$, we would have $\lim_{p\to\infty} f_p = 1$ uniformly on E_1. Choose $p_4 > 2p_3$ such that $f_{p_4} < 1 + \epsilon_4$ on E_1. Let $0 < b_4 < b_3$ and choose $a_4 > 0$ so small compared to b_4 that we still have $f_{p_4} < 1 + \epsilon_4$ on E_1. In particular $f_{p_4} < 1 + \epsilon_4$ on C.

Continuing this way, we see that $f_{p_{2n+1}} > 3/2 - \epsilon_{2n+1}$ on C, and $f_{p_{2n}} < 1 + \epsilon_{2n}$ on C. It follows that

$$\overline{\lim_{p\to\infty}} f_p \geq 3/2 \text{ on } C$$

and

$$\lim_{p\to\infty} f_p \leq 1 \text{ on } C.$$

BIBLIOGRAPHY

1. C. K. Chui, P. W. Smith, and J. D. Ward, "Favard's solution in the limit of W^k, p-splines", Trans. Amer. Math. Soc. 220 (1976), 299-305.

2. R. B. Darst, "Convergence of L_p approximations as $p\to\infty$", Proc. Amer. Math. Soc. 81 (1981), 433-436.

3. R. B. Darst, D. A. Legg, and D. W. Townsend, "The Polya algorithm in L_∞ approximation", Jour. Approx. Theory 38 (1983), 209-220.

4. R. B. Darst and S. Sahab, "Approximation of continuous and quasicontinuous functions by monotone functions", Jour. Approx. Theory 38 (1983), 9-27.

5. D. A. Legg and D. W. Townsend, "Best Monotone Approximation in $L_\infty[0,1]$", submitted, Jour. Approx. Theory.

6. J. Rice, The Approximation of Functions, Vol. I, Addison-Wesley, Reading, Mass., 1964.

7. J. Rice, The Approximation of Functions, Vol. II, Addison-Wesley, Reading, Mass., 1969.

8. V. A. Ubhaya, "Isotone Optimization II", Jour. Approx. Theory 12 (1974), 315-331.

DEPARTMENT OF MATHEMATICS
INDIANA UNIVERSITY-PURDUE UNIVERSITY AT FORT WAYNE
FORT WAYNE, IN 46805

Contemporary Mathematics
Volume 42, 1985

REPRESENTATION OF LATTICES AND EXTENSION OF MEASURES

Fon-che Liu[1]

ABSTRACT. Let L be a distributive lattice with a smallest element. Then L has a representation as lattice of closed subsets of a compact Hausdorff space S . When L is a boolean algebra, this representation can be used to extend the valuations of bounded variation on L to Radon measures on the Borel algebra of S .

1. Let L be a distributive lattice with the smallest element Θ . It is shown in [4] that L can be embedded as a lattice in a conditionally complete vector lattice without recourse to Stone Representation Theorem. The embedding is accomplished as follows. We shall understand by a valuation on L a real-valued function μ on L such that $\mu(\Theta) = 0$ and $\mu(x) + \mu(y) = \mu(x \wedge y) + \mu(x \vee y)\ \forall\ x,y \in L$. A valuation is said to be of bounded variation if for each $x \in L$

$$\sup \sum_{i=1}^{n-1} |\mu(x_{i+1}) - \mu(x_i)| < +\infty ,$$

where the supremum is taken over all finite chains $\Theta \leq x_1 \leq \dots \leq x_n$ in $[\Theta,x]$. Let X be the real vector space of all valuations of bounded variation on L, then it is known that X is a vector lattice with the set X_+ of all monotone nondecreasing valuations as its positive cond [1, p.240]. Then L is embedded in the order dual X^π of X through the mapping τ defined by

$$\tau(x)(\mu) = \mu(x) , \quad x \in L,\ \mu \in X.$$

We shall henceforth denote $\tau(x)$ simply by x^* for $x \in L$ and we shall not distinguish between L and L* which is the image of L under τ. Since X^π is the order dual of X, it is conditionally complete.

In this note we assume that L has also the largest element U and hence a norm $\|\cdot\|$ can be defined for X by

---

1980 Mathematics Subject Classification. 06B15, 06E15, 28A60.

[1]Partly supported by the National Science Council, Taipei.

$$\|\mu\| = |\mu|(U)\ , \quad \mu \in X\ ,$$

where as usual $|\mu| = \mu^+ + \mu^-$ with $\mu^+ = \mu \vee 0$ and $\mu^- = (-\mu) \vee 0$. Equipped with this norm X is a Banach lattice and, in fact, X and its topological dual X^* are an abstract (L)-space and an abstract (M)-space respectively in the sense of Kakutani. We note that as a set X^* is the same as X^π, that $[-U^*,U^*]$ is the closed unit ball of X^*, and that X^* is isometric and lattice isomorphic as Banach lattice with $C(S)$ of all continuous real-valued functions on a compact Hausdorff space S. Actually S can be taken as the set of all extremal points of the positive unit ball's surface $E^{**} = \{\eta \in X^{**} : \eta > 0 \text{ and } \|\eta\| = 1\}$ of X^{**} with the w*-topology. In this note we shall take S as such. For these facts we refer to [2], [3] and [5]. We have remarked in [4] that elements of X can be represented as Radon measures on S. Here we shall prove that L can be represented as a lattice of closed subsets of S and that when L is a boolean algebra, this representation renders possible the simultaneous extensions of all elements of X to Radon measures on the σ-algebra of Borel subsets of S, X being identified as the space of all finitely additive measures of bounded variation on an algebra of closed and open subsets of S which represents L.

2. This section is devoted to the representation of L as a lattice of closed subsets of S. Since the natural map from X^* into X^{***} is a lattice isomorphism, finite lattice operations in X^* may be carried out as if in X^{***}. We shall use this fact freely without mentioning it again. Let E^* and E^{**} be the positive unit surfaces of X^* and X^{**} respectively. We recall that S is the set of all extremal points of E^{**} equipped with w*-topology, with respect to which it is Hausdorff compact. For $\mu^* \in E^*$, let $T(\mu^*) = \{\nu \in S : \langle \mu^*, \nu \rangle = 1\}$, where $\langle \cdot , \cdot \rangle$ is the pairing between X^* and X^{**},

PROPOSITION 1. <u>For</u> $\mu^* \in E^*$, $T(\mu^*)$ <u>is a non-empty closed subset of</u> S.

Proof. Since S is equipped with w*-topology with respect to which it is Hausdorff compact [3], it is obvious that $T(\mu^*)$ is closed. Let $A = \{\nu \in E^{**} : \langle \mu^*, \nu \rangle = 1\}$. Since X^{**} is conditionally complete, it is clear that A is a non-empty convex w*-closed subset of E^{**}. By the Alaoglu-Bourbaki theorem and the Krein-Milman theorem the set $e(A)$ of all extremal points of A is non-empty. If A is an extremal subset of E^{**}, then $e(A) \subset S$ and the proof is established. Hence it remains to show that A is an extremal subset of E^{**}. For this purpose let ν_1, ν_2 be elements of E^{**} such that $\alpha\nu_1 + \beta\nu_2$ is in A for some $\alpha, \beta > 0$ with $\alpha + \beta = 1$. Then $\langle \mu^*, \alpha\nu_1 + \beta\nu_2 \rangle = 1 = \alpha\langle \mu^*, \nu_1 \rangle + \beta\langle \mu^*, \nu_2 \rangle$, $\langle \mu^*, \nu_1 \rangle \leq 1$, and $\langle \mu^*, \nu_2 \rangle \leq 1$ imply that $\langle \mu^*, \nu_i \rangle = 1$ for $i = 1,2$. Thus ν_1 and ν_2 are in A and hence A is an extremal subset of E^{**}. q.e.d.

PROPOSITION 2. $T(\mu_1^* \vee \mu_2^*) = T(\mu_1^*) \cup T(\mu_2^*)$ for $\mu_1^*, \mu_2^* \in E^*$; furthermore if $\mu_1^* \wedge \mu_2^* \in E^*$, then $T(\mu_1^* \wedge \mu_2^*) = T(\mu_1^*) \cap T(\mu_2^*)$.

Proof. Obviously, $T(\mu_1^* \vee \mu_2^*) \supset T(\mu_1^*) \cup T(\mu_2^*)$. Let $\nu \in T(\mu_1^* \vee \mu_2^*)$, we will show that $\nu \in T(\mu_1^*) \cup T(\mu_2^*)$. Suppose the contrary, then $\langle \mu_1^*, \nu \rangle < 1$ and $\langle \mu_2^*, \nu \rangle < 1$ from which we have

$$\begin{aligned}\langle \mu_1^* \vee \mu_2^*, \nu \rangle &= \sup\{\langle \mu_1^*, \nu_1 \rangle + \langle \mu_2^*, \nu_2 \rangle : \nu_1, \nu_2 \geq 0,\ \nu_1 + \nu_2 = \nu\} \\ &= \sup\{\langle \mu_1^*, \alpha\nu \rangle + \langle \mu_2^*, \beta\nu \rangle : \alpha \geq 0,\ \beta \geq 0,\ \alpha + \beta = 1\} \\ &= \max\{\langle \mu_1^*, \nu \rangle, \langle \mu_2^*, \nu \rangle\} < 1,\end{aligned}$$

which contradicts the fact that $\nu \in T(\mu_1^* \vee \mu_2^*)$. In the above we have used the fact that ν is an extremal point of E^{**} so that $\nu_1, \nu_2 \geq 0$ and $\nu_1 + \nu_2 = \nu$ imply $\nu_1 = \alpha\nu$, $\nu_2 = \beta\nu$, $\alpha, \beta \geq 0$. We have shown that

$$T(\mu_1^* \vee \mu_2^*) = T(\mu_1^*) \cup T(\mu_2^*).$$

Now suppose that $\mu_1^* \wedge \mu_2^* \in E^*$. It is clear that $T(\mu_1^* \wedge \mu_2^*) \subset T(\mu_1^*) \cap T(\mu_2^*)$. Let $\nu \in T(\mu_1^*) \cap T(\mu_2^*)$, then $\langle \mu_i^*, \nu \rangle = 1$, $i = 1,2$. But

$$\langle \mu_1^* \wedge \mu_2^*, \nu \rangle = \mathrm{Inf}\{\langle \mu_1^*, \nu_1 \rangle + \langle \mu_2^*, \nu_2 \rangle : \nu_1, \nu_2 \geq 0,\ \nu_1 + \nu_2 = \nu\} = 1,$$

where the last equality also follows from the fact that ν is an extremal point of E^{**} as in the first paragraph of the proof. Thus $\nu \in T(\mu_1^* \wedge \mu_2^*)$. q.e.d.

For convenience let us extend T so that it is also defined at 0^*, the origin of X^*, by simply assigning the empty subset of S to $T(0^*)$.

Since for $x \in L$, $x \neq \Theta$, it is readily verified that $x^* \in E^*$, by restricting T to L^* we obtain the following theorem from Propositions 1 and 2:

THEOREM 1. The map $x \to T(x^*)$, $x \in L$, is a lattice isomorphism of L with a lattice $\mathcal{A}(L)$ of closed subsets of S. Furthermore if L is a boolean algebra, then $\mathcal{A}(L)$ is an algebra of closed and open subsets of S and the map $x \to T(x^*)$, $x \in L$, also preserves complementation.

Proof. We need only show that the map $x \to T(x^*)$, $x \in L$, is one to one. For this we remark that for $x, y \in L$, $x \neq y$, there is $\mu \in X$ with $\|\mu\| = 1$ and $\{\mu(x), \mu(y)\} = \{0,1\}$ [4]. q.e.d.

3. Up until now our presentation does not depend on the representation of X^* by $C(S)$ as Banach lattice; now we invoke this fact to consider the extension of finitely additive measures to σ-additive measures. We consider the case that L is a boolean algebra. Then $\mathcal{A}(L)$ is an algebra of closed and open subsets of S. Elements of X can be considered as finitely additive measures of bounded variation on $\mathcal{A}(L)$. By abuse of notation we use the same notation for elements of X when they are considered as measures on $\mathcal{A}(L)$, that is, for $\mu \in X$

$$\mu[T(x^*)] = \mu(x), \quad x \in L.$$

Since elements of X are continuous linear functionals on X^*, by Kakutani's representation of X^* by $C(S)$ as Banach lattice, to each element μ of X, there is a unique Radon measure $\hat{\mu}$ on S such that

$$\langle \nu^*, \mu \rangle = \int_S \langle \nu^*, s \rangle d\hat{\mu}(s) \quad \forall\ \nu^* \in X^*;$$

in particular, we have

$$\mu(x) = \int_S \langle x^*, s \rangle\, d\hat{\mu}(s), \quad x \in L.$$

But for $x \in L$, $x + x^c = x \vee x^c + x \wedge x^c = U$, hence $\langle x^*, s \rangle = 1$ for $s \in T(x^*)$ and $\langle x^*, s \rangle = 0$ for $s \in S \setminus T(x^*)$, where x^c is the complement of x in L. Thus,

$$\mu(x) = \hat{\mu}[T(x^*)] \quad \forall\ x \in L.$$

Consequently, we have the following theorem:

THEOREM 2. For any boolean algebra L, there is a compact Hausdorff space S such that L can be represented as an algebra $\mathcal{A}(L)$ of closed and open subsets of S and such that each valuation of bounded variation on L can be represented as a restriction of a Radon measure on S to $\mathcal{A}(L)$.

COROLLARY. Let L be an algebra of subsets of a given set. Then L can be embedded as an algebra $\mathcal{A}(L)$ of closed and open subsets of a compact Hausdorff space S such that each finitely additive measure of bounded variation on $\mathcal{A}(L)$ can be extended to a Radon measure on the Borel field of S. In particular, any finitely additive probability measure on $\mathcal{A}(L)$ can be extended to be a Radon measure on S.

BIBLIOGRAPHY

1. G. Binkhoff, "Lattice Theory", Amer. Math. Soc. Colloquium Pub., 1967.

2. M.M. Day, Normed Linear Spaces, Chapter 2, 3rd ed., Springer-Verlag, 1973.

3. S. Kakutani, "Concrete Representation of Abstract (M)-Spaces", Annals of Math., 42 (1941), 994-1024.

4. F.C. Liu, "Embedding Distributive Lattices in Vector Lattices", Bull. Inst. Math., Academia Sinica-Taipei, to appear in September 1983.

5. H.H. Schaefer, Banach Lattices and Positive Operators, Chapter 2, Springer-Verlag, 1974.

ACADEMIA SINICA and NATIONAL TAIWAN UNIVERSITY
TAIPEI, TAIWAN

Contemporary Mathematics
Volume 42, 1985

TRANSFORMATION AND MULTIPLICATION OF DERIVATIVES

Jan Mařík

ABSTRACT. The author investigates homeomorphisms h of the interval [0,1] onto itself such that the composite function $f \circ h$ belongs to a certain class for each f belonging to another class. These classes consist of derivatives or approximately continuous functions. Then he investigates functions g such that the product fg is a derivative for each derivative f or each Lebesgue function f and shows a connection between families of such functions h and g.

0. INTRODUCTION. Let H be the family of all increasing homeomorphisms of the interval [0,1] onto itself. A. M. Bruckner characterized in [1] the class of all $h \in H$ such that the composite function $f \circ h$ is approximately continuous for each approximately continuous f. In [2] he investigated homeomorphisms $h \in H$ such that $f \circ h \in D$ for each $f \in D$ for which f^2 also belongs to D, where D is the class of all derivatives. M. Laczkovich and G. Petruska characterized in [5] the smaller class of all $h \in H$ such that $f \circ h \in D$ for all $f \in D$. R. J. Fleissner described in [4] the system of all functions g such that $fg \in D$ for each $f \in D$. The present paper contains, among other things, improvements of some of the results obtained in the mentioned articles and also shows connections between these results.

The word "function" means, throughout the paper, a mapping of a subset of $R = (-\infty, \infty)$ to $R^* = R \cup \{-\infty, \infty\}$. A function whose range is in R is called a finite function. Multiplication in R^* is defined in the usual way; in particular, $a\infty = \infty$ for $a > 0$. The word "continuous" refers to the usual topology in R; thus, "continuous function" always means a finite function.

The letter J stands for the interval [0,1]. The system of all finite [bounded] derivatives on J is denoted by D [bD]. The

1980 Mathematics Subject Classification. 26A24, 26A15.

system of all finite functions that are approximately continuous with respect to J at each point of J is denoted by C_{ap}; the meaning of bC_{ap} is obvious.

Symbols like $\int_a^b f$, $\int_a^b f(t)dt$ denote the corresponding Lebesgue or Perron integral. If $a,b \in R$, $a<b$ and if $K=[a,b]$, we write also $\int_K f$. The meaning of expressions "L-integrable," "P-integrable" is obvious.

The outer Lebesgue measure of a set $S \subset R$ is denoted by $|S|$. Words like "measurable" always refer to the Lebesgue measure in R.

A finite function f on J belongs to D if and only if $f(a) = \lim (x-a)^{-1} \int_a^x f$ $(x \to a, x \in J)$ for each $a \in J$. A finite function f on J is called a Lebesgue function if and only if $(x-a)^{-1} \int_a^x |f-f(a)| \to 0$ $(x \to a, x \in J)$ for each $a \in J$. The system of all Lebesgue functions is denoted by L. It is easy to see that L is a vector space, $bC_{ap} \subset L \subset D$ and that each Lebesgue function is L-integrable. The system of all functions $f \in D$ such that f^2 also belongs to D is denoted by W.

1. DERIVATIVES AND APPROXIMATELY CONTINUOUS FUNCTIONS. This part contains some simple assertions that will be used later. The proofs of 1.1 and 1.2 are left to the reader.

<u>1.1</u>. Let $f \in D$, $f \geq 0$, $c \in J$, $f(c) = 0$. Then f is approximately continuous at c with respect to J.

<u>1.2</u>. Let $f \in L$. Let g be a finite function on R such that $|g(y)-g(x)| \leq |y-x|$ for any $x,y \in R$. Then $g \circ f \in L$.

<u>1.3</u>. $L \subset C_{ap}$.

Proof. Let $c \in J$. By 1.2 we have $|f-f(c)| \in D$. Now we apply 1.1.

<u>1.4</u>. $W \subset L$. (See [2], Theorem 1.)

Proof. Let $w \in W$, $c \in J$, $g = w-w(c)$. It follows from the Cauchy inequality and from the relation $g^2 \in D$ that $((x-c)^{-1} \int_c^x |g|)^2 \leq (x-c)^{-1} \int_c^x g^2 \to 0$ $(x \to c, x \in J)$.

<u>1.5</u>. Let $f \in D$ and let $f \vee a \in D$ for each $a \in R$. Then $f \in L$.

Proof. Let $c \in J$, $a = f(c)$. Obviously $f + a = (f \vee a) + (f \wedge a)$ whence $f \wedge a \in D$. Since $|f-a| = (f \vee a) - (f \wedge a)$, we have $|f-a| \in D$. Thus $(x-c)^{-1} \int_c^x |f-f(c)| \to 0 (x \to c, x \in J)$, $f \in L$.

<u>1.6</u>. Let g be a finite function on J such that $g \vee b \in D$ for each $b \in R$. Then $g \in C_{ap}$.

Proof. It follows from 1.5 and 1.3 that $g \vee c \in C_{ap}$ for each $c \in R$. Hence $g \in C_{ap}$.

1.7. Let f be a finite function such that $(f \wedge a) \vee b \in D$ for any $a,b \in R$. Then $f \in C_{ap}$. (See [3], p.50, Theorem 4.1.)

(This follows from 1.6.)

1.8. Let $f,g \in C_{ap}$, $f \in D$ and $|g| \leq f$. Then $g \in L$.

Proof. We may suppose that $g \geq 0$. Let $c \in J$, $a = f(c)$, $f_1 = f \wedge a$, $g_1 = g \wedge a$, $f_2 = f - f_1$, $g_2 = g - g_1$. Then $f_2 = (f \vee a) - a$, $g_2 = (g \vee a) - a$. Since $f_1 \in bC_{ap}$, we have $f_1, f_2 \in D$. Obviously $0 \leq g_2 \leq f_2$, $(x-c)^{-1} \int_c^x g_2 \leq (x-c)^{-1} \int_c^x f_2 \to f_2(c) = 0$ $(x \to c,\ x \in J)$. Since $g_1 \in bC_{ap}$, we have $g_1 \in L$. Thus $g = g_1 + g_2 \in L$.

1.9. W is a vector space. If $f,g \in W$, then $fg \in L$.

Proof. Let $f,g \in W$. It follows from 1.4 and 1.3 that $W \subset C_{ap}$. Since $2|fg| \leq f^2 + g^2$, we have, by 1.8, $fg \in L$. Now it is obvious that $(f+g)^2 \in D$ which shows that W is a vector space.

1.10. Let $f \in L$, $f \geq 0$. Then $f^{1/2} \in W$.

Proof. Obviously $f^{1/2} < 1+f$. Now we apply 1.3 and 1.8.

1.11. Let $f \in L$. Then there are $v,w \in W$ such that $f = vw$.

Proof. By 1.2 and 1.10 there are $w_1, w_2 \in W$ such that $w_1^2 = f \vee 0$, $w_2^2 = (-f) \vee 0$. Set $v = w_1 + w_2$, $w = w_1 - w_2$. By 1.9 we have $v,w \in W$. Obviously $vw = f$.

2. INTEGRATION AND COMPOSITION. This part is connected with the problem of finding conditions under which a composite function is a derivative. Most of the results are of auxiliary character. However, sections 2.3 and 2.4 contain simple estimates of integrals and may be of independent interest.

Let $a,b \in R$, $a < b$ and let φ be a function on the interval $K = [a,b]$. If $\varphi(x) = \pm\infty$ for some $x \in K$, we set $\mathrm{var}(\varphi,K) = \infty$. If $\varphi(K) \subset R$, we define, as usual, $\mathrm{var}(\varphi,K)$ as the least upper bound of the set of all sums $\Sigma_{j=1}^{n} |\varphi(t_j) - \varphi(t_{j-1})|$, where $a = t_0 < t_1 < \ldots < t_n = b$. We write $\mathrm{var}(\varphi,K) = \mathrm{var}(\varphi,a,b) = \mathrm{var}(\varphi,b,a)$. Further we set $\sup(\varphi,a,b) = \sup(\varphi,b,a) = \sup \varphi(K)$ and $\mathrm{var}(\varphi,c,c) = 0$, $\sup(\varphi,c,c) = \varphi(c)$ for each $c \in K$.

If $\varphi(K) \subset R$, then $U^+\varphi(a)$ $[L^+\varphi(a)]$ is the right upper [lower] derivate of φ at a; the meaning of $U^-\varphi(x)$, $L^-\varphi(x)$ (for $x \in (a,b]$), $U\varphi(x)$, $L\varphi(x)$ (for $x \in (a,b)$) is obvious. If $[a,b]$ is the domain of definition of φ, we write sometimes $U\varphi(a)$, $\varphi'(a)$ instead of $U^+\varphi(a)$, $\varphi'^+(a)$ etc.

In 2.1 and 2.2 we state without proof two well-known results of integration theory. (See, e.g., [6], Chapter VIII.) Symbols a,b,K have the same meaning as before.

2.1. Let f be P-integrable on K and let φ be a finite non-negative nonincreasing function on K. Then $f\varphi$ is P-integrable on K and there is a $\xi \in K$ such that $\int_K f\varphi = \varphi(a) \int_a^\xi f$.

2.2. Let f be a function on K that is P-integrable on $[x,b]$ for each $x \in (a,b)$. If $\lim_{x\to a+} \int_x^b f = \lambda \in R$, then f is P-integrable on K and $\int_K f = \lambda$.

2.3. Let $\sigma \in R$. Let φ be a nonnegative nonincreasing function on $(a,b]$ such that $\int_K \varphi < \infty$. Let f be a function P-integrable on K such that $|\int_a^x f| \le \sigma(x-a)$ for each $x \in K$. Then the function $f\varphi$ is P-integrable on K and $|\int_K f\varphi| \le \sigma \int_K \varphi$.

Proof. Let $\alpha, \beta \in (a,b]$, $\alpha < \beta$. Set $f_1 = f - \alpha$. By 2.1 there is a $\xi \in [\alpha,\beta]$ such that $\int_\alpha^\beta f_1\varphi = \varphi(\alpha) \int_\alpha^\xi f_1 = \varphi(\alpha)(\int_a^\xi f_1 - \int_a^\alpha f_1)$. Since $\int_a^\xi f_1 \le 0$, we have $\int_\alpha^\beta f\varphi \le M$, where $M = \sigma \int_\alpha^\beta \varphi + 2\sigma\varphi(\alpha)(\alpha - a)$. Taking $-f$ instead of f we see that $|\int_\alpha^\beta f\varphi| \le M$. Since $(\alpha-a)\varphi(\alpha) \le \int_a^\alpha \varphi$, there is a finite limit $\lambda = \lim \int_x^b f\varphi$ $(x \to a+)$ and $|\lambda| \le \sigma \int_K \varphi$. By 2.2 we have $\lambda = \int_K f\varphi$.

2.4. Let $\sigma, A \in R$. Let f be a P-integrable function on K such that $|\int_a^x f| \le \sigma(x-a)$ for each $x \in K$. Let α be a function on K such that $\int_K \operatorname{var}(\alpha,t,b)\,dt = A$. Then $f\alpha$ is P-integrable on K and

$$|\int_K f\alpha| \le \sigma(2A + |\int_K \alpha|).$$

Proof. Suppose first that $\alpha(b) = 0$. Set $\psi(x) = \operatorname{var}(\alpha,x,b)$ $(x \in (a,b])$, $\alpha_1 = (\psi + \alpha)/2$, $\alpha_2 = (\psi - \alpha)/2$. By 2.3 we have $|\int_K f\alpha_j| \le \sigma \int_K \alpha_j$ for $j = 1,2$. Hence

$$|\int_K f\alpha| \le \sigma A. \tag{1}$$

In the general case we have $|K|\alpha(b) = \int_K(\alpha(b) - \alpha) + \int_K \alpha$, $\int_K f\alpha = \int_K f\cdot(\alpha - \alpha(b)) + \alpha(b)\int_K f$ so that $|K||\alpha(b)| \le A + |\int_K \alpha|$ and (see (1)) $|\int_K f\alpha| \le \sigma A + |\alpha(b)|\sigma|K| \le \sigma(2A + |\int_K \alpha|)$.

2.5. Let f be L-integrable on K. Let γ be a nonnegative measurable function on K such that $0 < \int_a^x \gamma < \infty$ for each $x \in (a,b)$. Suppose that $(x-a)^{-1} \int_a^x |f - f(a)| \to 0$ and that

$$\limsup (\int_a^x \gamma)^{-1} \int_a^x \sup(\gamma,t,x)\,dt < \infty \qquad (x \to a+). \tag{2}$$

Then there is a $c \in (a,b)$ such that $f\gamma$ is L-integrable on $[a,c]$ and we have

$$(\int_a^x \gamma)^{-1} \int_a^x f\gamma \to f(a) \qquad (x \to a+).$$

Proof. We may suppose that $f(a) = 0$. For each $x \in (a,b]$ set $\sigma(x) = \sup\{(t-a)^{-1}\int_a^t |f|;\ a < t \le x\}$. Obviously $\sigma(a+) = 0$. There is a $c \in (a,b)$ such that $\int_a^c \sup(\gamma,t,c)\,dt < \infty$. Choose an $x \in (a,c]$ and set $\varphi(t) = \sup(\gamma,t,x)$ $(a < t \le x)$. By 2.3,

$\int_a^x |f|\varphi \le \sigma(x)\int_a^x \varphi$. Since $|f\gamma| \le |f|\varphi$ on $(a,x]$, we have (see (2)) $(\int_a^x \gamma)^{-1}\int_a^x f\gamma \to 0$ $(x \to a+)$.

2.6. Let α,φ,f be P-integrable functions on K and let $\lambda \in R$. Suppose that $\int_a^x \varphi > 0$ for each $x \in (a,b)$, $(\int_a^x \varphi)^{-1}\int_a^x \alpha \to \lambda$, $(x-a)^{-1}\int_a^x f \to f(a)$ and that

(3) $$\limsup (\int_a^x \varphi)^{-1}\int_a^x \mathrm{var}(\alpha,t,x)dt < \infty \qquad (x \to a+).$$

Then there is a $c \in (a,b)$ such that the function $f\alpha$ is P-integrable on $[a,c]$ and we have

(4) $$(\int_a^x \varphi)^{-1}\int_a^x f\alpha \to \lambda f(a) \qquad (x \to a+).$$

Proof. We may suppose that $f(a) = 0$. For each $x \in (a,b)$ set $\sigma(x) = \sup\{(t-a)^{-1}|\int_a^t f|;\ a < t \le x\}$. Obviously $\sigma(a+) = 0$. There is a $c \in (a,b)$ such that $\int_a^c \mathrm{var}(\alpha,t,c)dt < \infty$. By 2.4 we have $|\int_a^x f\alpha| \le \sigma(x)(|\int_a^x \alpha| + 2\int_a^x \mathrm{var}(\alpha,t,x)dt)$ for each $x \in (a,c]$. This easily implies (4).

2.7. Let $b \in (0,\infty)$, $K = [0,b]$. Let g be a continuous increasing function on K, $h = g^{-1}$. Set $\gamma(b) = U^-g(b)$, $\gamma = Ug$ on $(0,b)$. Let M be a number less that $\int_K \sup(\gamma,t,b)dt$. Then there exist an $a \in (0,b)$ and a nonnegative piecewise linear function f on K such that $f = 0$ on $(0,a) \cup \{b\}$, $\int_0^x f \le x$ for each $x \in K$ and

$$\int_{g(K)} f\circ h > M.$$

Proof. For each $x \in (0,b]$ define $\varphi(x) = \sup(\gamma,x,b)$.

Suppose first that $\varphi(c) = \infty$ for some $c \in (0,b)$. Set $A = 2|M|/c$. There is a $z \in [c,b]$ such that $\gamma(z) > A$. There are numbers p,q such that either p=z or q=z, $q/2 < p < q \le b$ and $g(q) - g(p) > A(q-p)$. There are numbers α,β such that $p < \alpha < \beta < q$ and $g(\beta) - g(\alpha) > A(q-p)$. Let f be a function such that $f = 0$ on $[0,p]$ and on $[q,b]$ (which means $\{b\}$ for q=b), $f = p/(q-p)$ on $[\alpha,\beta]$ and that f is linear on $[p,\alpha]$ and on $[\beta,q]$. Since $\int_0^p f = 0$ and $\int_K f < p$, we have $\int_0^x f \le x$ for each $x \in K$. Since $2p > q \ge z \ge c$, we have $Ap \ge M$. Obviously $\int_{g(\alpha)}^{g(\beta)} f\circ h = (g(\beta)-g(\alpha))p/(q-p) > Ap$ so that $\int_{g(K)} f\circ h > M$.

Now suppose that $\varphi((0,b)) \subset R$. There are numbers $c \in (0,b)$, $Q \in (1,\infty)$ and $\varepsilon \in (0,\infty)$ such that $\int_c^b \varphi > QM + b\varepsilon$. There are $t_0,\dots,t_n$ such that $c = t_0 < t_1 < \dots < t_n = b$ and that $t_i < Qt_{i-1}$ for $i = 1,\dots,n$. There are integers s and j_k $(s \ge 1)$ such that $\varphi(t_0) = \varphi(t_{j_1}-1) > \varphi(t_{j_1}) = \varphi(t_{j_2-1}) > \varphi(t_{j_2}) = \dots > \varphi(t_{j_{s-1}}) = \varphi(t_{n-1})$. Set $j_0 = 0$, $j_s = n$, $v_k = t_{j_k}$ $(k = 0,\dots,s)$, $A_k = \varphi(t_{j_k-1}) - \varepsilon$

$(k=1,\dots,s)$. For $k=1,\dots,s-1$ there is a $z_k \in [t_{j_k-1}, t_{j_k}]$ such that $\gamma(z_k) > A_k$; there is a $z_s \in [t_{n-1}, t_n]$ such that $\gamma(z_s) > A_s$. Note that $v_k < Qz_k$. There are a_k, b_k such that either $a_k = z_k$ or $b_k = z_k$, $a_1 < b_1 < \dots < a_s < b_s \leq b$, $v_k < Qa_k$ and $g(b_k) - g(a_k) > A_k(b_k - a_k)$ for $k=1,\dots,s$. There are numbers α_k, β_k such that $a_k < \alpha_k < \beta_k < b_k$ and that $g(\beta_k) - g(\alpha_k) > A_k(b_k - a_k)$. Let F be a function on $[0,b]$ such that $F=0$ on $[0,a_1] \cup [b_s,b]$ and on $[b_k, a_{k+1}]$ for $k=1,\dots,s-1$, $F = (v_k - v_{k-1})/(b_k - a_k)$ on $[\alpha_k, \beta_k]$ and that F is linear on each of the intervals $[a_k, \alpha_k]$ and $[\beta_k, b_k]$ for $k=1,\dots,s$. Set $a_{s+1} = b$. Obviously $\int_{a_k}^{a_{k+1}} F < v_k - v_{k-1}$. If $a_k < x \leq a_{k+1}$, then $\int_0^x F < v_k < Qa_k < Qx$. Further $\int_{g(\alpha_k)}^{g(\beta_k)} f \circ h = (g(\beta_k) - g(\alpha_k))(v_k - v_{k-1})/(b_k - a_k) > A_k(v_k - v_{k-1})$. Since $A_k = \varphi(t_i) - \varepsilon$ for $i = j_{k-1}, \dots, j_k - 1$, we have $A_k(v_k - v_{k-1}) = \Sigma_{i=j_{k-1}+1}^{j_k} \varphi(t_{i-1})(t_i - t_{i-1}) - \varepsilon(v_k - v_{k-1}) \geq \int_{v_{k-1}}^{v_k} \varphi - \varepsilon(v_k - v_{k-1})$. Hence $\int_{g(K)} F \circ h > \int_{v_0}^{v_s} \varphi - \varepsilon b > QM$. Now we set $f = F/Q$.

2.8. Let $a,b \in R$, $a<b$, $K=[a,b]$. Let g be a continuous increasing function on K and let $A,B \in R$. If $Ug \geq A$ on (a,b), then $g(b)-g(a) \geq A|K|$; if $Lg \leq B$ on (a,b), then $g(b)-g(a) \leq B|K|$.

Proof. If $Ug \geq A$ on (a,b), then, by Theorem 7.4 of Chapter IV of [6], we have $g(b)-g(a) \geq A|K|$.

Now let $Lg \leq B$ on (a,b) and let N be the set of all points $x \in (a,b)$ for which $g'(x)$ does not exist. By Theorems 4.4 and 6.5 of Chapter VII of [6] we have $|g(N)| = 0$ and $|g(a,b) \setminus N| \leq B|K|$. Therefore $g(b)-g(a) = |g((a,b))| \leq B|K|$.

2.9. Let g be a continuous increasing function on $[a,b]$. Then

$$\liminf Ug(x) \leq L^+g(a), \quad U^+h(a) \leq \limsup Lg(x) \qquad (x \to a+).$$

(This follows at once from 2.8.)

2.10. Let $b \in (0,\infty)$. Let ψ be a finite nonincreasing function on $(0,b]$ with $\psi(b) = 0$ and let Q be a number less than $\int_0^b \psi$. Then there are numbers $t_0,\dots,t_n$ such that $0 < t_0 < \dots < t_n < b$, ψ is continuous at t_j for $j=1,\dots,n$ and

$$\Sigma_{j=1}^{n} (t_j - t_{j-1}) \psi(t_j) > Q. \tag{5}$$

Proof. There is a $t_0 \in (0,b)$ and an integer $n > 1$ such that

$$\int_{t_0}^b \psi > Q + 2b\psi(t_0)/n. \tag{6}$$

Let $z_j = t_0 + j(b-t_0)/n$ $(j = 0,\ldots,n)$. There are $t_j \in (z_{j-1}, z_j)$ such that ψ is continuous at t_j $(j = 1,\ldots,n)$. Set $t_{n+1} = b$, $S = \Sigma_{j=1}^{n}(t_j - t_{j-1})\psi(t_j)$, $T = \Sigma_{j=1}^{n+1}(t_j - t_{j-1})\psi(t_{j-1})$. Obviously $\int_{t_0}^{b}\psi \le T$, $t_j - t_{j-1} < 2b/n$ so that $\int_{t_0}^{b}\psi - S \le T - S = \Sigma_{j=1}^{n+1}(t_j - t_{j-1})(\psi(t_{j-1}) - \psi(t_j)) \le 2b\psi(t_0)/n$. This and (6) proves (5).

<u>2.11</u>. Let $a,b \in R$, $K = [a,b]$. Let g be a continuous increasing function on K, $h = g^{-1}$. Let γ be a function on K such that $L^+g(a) \le \gamma(a) \le U^+g(a)$, $L^-g(b) \le \gamma(b) \le U^-g(b)$ and that $Lg \le \gamma \le Ug$ on (a,b).

Let A be a number less than $\mathrm{var}(\gamma,K)$. Then there is a function f piecewise linear on K such that $f(a) = f(b) = \int_K f = 0$, $|\int_a^x f| \le 1$ for each $x \in K$ and that

$$\int_{g(K)} f\circ h > A .$$

Proof. Suppose first that there is a $c \in K$ such that $\gamma(c) = \infty$. Let, e.g., $U^+g(c) = \infty$. Set $s = (c+b)/2$. Let F be a nonnegative piecewise linear function on $[s,b]$ such that $F(s) = F(b) = 0$ and $\int_s^b F = 1$. Set $B = \int_{g(s)}^{g(b)} F\circ h$. There is a $b_0 \in (c,s)$ such that $\dfrac{g(b_0)-g(c)}{b_0-c} > A + B$. There is a $\delta \in (0,\infty)$ such that $c + 2\delta < b_0$ and that $(g(b_0-\delta)-g(c+\delta))/(b_0-c-\delta) > A + B$. Let f be a function on K such that $f(t) = 0$, if $a \le t \le c$, $f = (b_0-c-\delta)^{-1}$ on $[c+\delta, b_0-\delta]$, $f = 0$ on $[b_0,s]$, $f = -F$ on $[s,b]$ and that f is linear on $[c,c+\delta]$ and on $[b_0-\delta,b_0]$. Then $f(a) = f(b) = \int_K f = 0$, $|\int_a^x f| \le 1$ for each $x \in K$ and $\int_{g(K)} f\circ h = \int_{g(c)}^{g(b_0)} f\circ h + \int_{g(s)}^{g(b)} f\circ h > A + B - B = A$.

Now suppose that $\gamma(K) \subset R$. Then there are numbers $t_0,\ldots,t_n$ and η such that $a = t_0 < t_1 < \ldots < t_n = b$, $\eta > 0$ and that

$$\Sigma_{j=1}^{n}|\gamma(t_j)-\gamma(t_{j-1})| > A + 2n\eta . \tag{7}$$

We may suppose that $(\gamma(t_j)-\gamma(t_{j-1}))\cdot(\gamma(t_{j+1})-\gamma(t_j)) < 0$ for $j = 1,\ldots,n-1$. Let, e.g., $\gamma(t_0) > \gamma(t_1) < \gamma(t_2) > \ldots$. Then

$$\Sigma_{j=1}^{n}|\gamma(t_j)-\gamma(t_{j-1})| = \tag{8}$$

$$\gamma(t_0) - 2\gamma(t_1) + \ldots + 2(-1)^{n-1}\gamma(t_{n-1}) + (-1)^n\gamma(t_n) .$$

Choose an $\varepsilon \in (0,\infty)$ such that $2\varepsilon < t_j - t_{j-1}$ for $j = 1,\ldots,n$. There are $z_j \in (a,b)$ such that $(g(z_j)-g(t_j))/(z_j-t_j) > \gamma(t_j) - \eta$ for j even, $(g(z_j)-g(t_j))/(z_j-t_j) < \gamma(t_j) + \eta$ for j odd and $|z_j-t_j| < \varepsilon$ for $j = 0,\ldots,n$.

If $z_j < t_j$, set $a_j = z_j$, $b_j = t_j$; if $z_j > t_j$, set $a_j = t_j$, $b_j = z_j$. Obviously $a = a_0 < b_0 < a_1 < \ldots < b_n = b$. There is a $\delta \in (0,\infty)$ such that $2\delta < b_j - a_j$ for $j = 0,\ldots,n$ and that $(g(b_j-\delta)-g(a_j+\delta))/(b_j-a_j-\delta) > \gamma(t_j)-\eta$ for j even, $(g(b_j)-g(a_j))/(b_j-a_j-\delta) < \gamma(t_j)+\eta$ for j odd. Now let $c_j = (b_j-a_j-\delta)^{-1}$ $(j=0,\ldots,n)$ and let f be a function on K such that $f = c_0$ on $[a_0+\delta, b_0-\delta]$, $f = 2(-1)^j c_j$ on $[a_j+\delta, b_j-\delta]$ $(j=1,\ldots,n-1)$, $f = (-1)^n c_n$ on $[a_n+\delta, b_n-\delta]$, $f = 0$ on $K \setminus \bigcup_{j=0}^n (a_j,b_j)$ and f is linear on each of the intervals $[a_j, a_j+\delta]$ and $[b_j-\delta, b_j]$. Set $s_0 = a$, $s_n = b$ and $s_j = (a_j+b_j)/2$ for $j=1,\ldots,n-1$. It is easy to see that $\int_{a_0}^{b_0} f = 1$, $\int_{a_n}^{b_n} f = (-1)^n$, $\int_{a_j}^{s_j} f = \int_{s_j}^{b_j} f = (-1)^j$ for $j=1,\ldots,n-1$, $\int_{s_{j-1}}^{s_j} f = 0$ for $j=1,\ldots,n$ and that $|\int_a^x f| \le 1$ for each $x \in K$. If j is even and $0<j<n$, then $\int_{g(a_j)}^{g(b_j)} f\circ h > 2c_j(g(b_j-\delta)-g(a_j+\delta)) > 2\gamma(t_j)-2\eta$; if j is odd and $0<j<n$, then $\int_{g(a_j)}^{g(b_j)} f\circ h > -2c_j(g(b_j)-g(a_j)) > -2\gamma(t_j)-2\eta$. Similarly, $\int_{g(a_0)}^{g(b_0)} f\circ h > \gamma(t_0)-\eta$ and $\int_{g(a_n)}^{g(b_n)} f\circ h > (-1)^n\gamma(t_n)-\eta$. Thus (see (7), (8)) $\int_{g(K)} f\circ h > \gamma(t_0) - 2\gamma(t_1) + \ldots + 2(-1)^{n-1}\gamma(t_{n-1}) + (-1)^n\gamma(t_n) - 2n\eta > A$.

2.12. Let $b \in (0,\infty)$, $K = [0,b]$. Let g be a continuous increasing function on K, $h = g^{-1}$. Let γ be a function on $(0,b]$ such that $L^- g(b) \le \gamma(b) \le U^- g(b)$ and $Lg \le \gamma \le Ug$ on $(0,b)$. Let M be a number less than $\int_K \mathrm{var}(\gamma,t,b)dt$. Then there is an $a \in (0,b)$ and a function f piecewise linear on K such that $f = 0$ on $(0,a)$, $f(b) = \int_K f = 0$, $|\int_0^x f| \le x$ for each $x \in K$ and $\int_{g(K)} f\circ h > M$.

Proof. For each $x \in (0,b]$ define $\psi(x) = \mathrm{var}(\gamma,x,b)$.

Suppose first that $\psi(c) = \infty$ for some $c \in (0,b)$. There is an $a \in (0,c)$ such that $g'(a)$ exists. Then, obviously, $\gamma(a) = g'(a) = L^+g(a) = U^+g(a)$ and $\mathrm{var}(\gamma,a,b) = \infty$. It follows from 2.11 that there is a piecewise linear function F on $[a,b]$ such that $F(a) = F(b) = \int_a^b F = 0$, $|\int_a^x F| \le 1$ for each $x \in (a,b)$ and that $a\int_{g(a)}^{g(b)} F\circ h > M$. It is easy to see that the function f defined by $f = 0$ on $[0,a]$ and $f = aF$ on $(a,b]$ satisfies our requirements.

Now suppose that $\psi((0,b]) \subset R$. Choose a number Q with $M < Q < \int_K \psi$, find numbers t_j according to 2.10 and set $t_{n+1} = b$. Then $Q < \Sigma_{j=1}^n (t_j - t_{j-1})\psi(t_j) = -t_0\psi(t_1) + \Sigma_{j=1}^n t_j(\psi(t_j) - \psi(t_{j+1})) \le \Sigma_{j=1}^n t_j \mathrm{var}(\gamma, t_j, t_{j+1})$. Since ψ is continuous at t_j, γ is continuous at t_j as well. It follows easily from 2.9 that $g'(t_j)$ exists for $j = 1,\dots,n$. Let $\varepsilon \in (0,(Q-M)/(nb))$. By 2.11 (note that $\gamma(t_1) = g'(t_1) = L^+g(t_1)$ etc.) there are functions f_j piecewise linear on $[t_j,t_{j+1}]$ such that $f_j(t_j) = f_j(t_{j+1}) = \int_{t_j}^{t_{j+1}} f_j = 0$, $|\int_{t_j}^{x} f_j| \leq 1$ for each $x \in [t_j,t_{j+1}]$ and that $\int_{g(t_j)}^{g(t_{j+1})} f_j \circ h > \mathrm{var}(\gamma,t_j,t_{j+1}) - \varepsilon$. Let f be a function on K such that $f = 0$ on $[0,t_1]$ and $f = t_j f_j$ on $[t_j,t_{j+1}]$ $(j = 1,\dots,n)$. Then $f(b) = \int_K f = 0$ and $\int_{g(K)} f \circ h > \Sigma_{j=1}^{n} t_j(\mathrm{var}(\gamma,t_j,t_{j+1}) - \varepsilon) > Q - nb\varepsilon > M$. If $t_j \leq x \leq t_{j+1}$, then $|\int_0^x f| = |\int_{t_j}^x f| \leq t_j \leq x$. This completes the proof.

3. TRANSFORMATIONS VIA INNER HOMEOMORPHISMS. Let AC be the system of all absolutely continuous functions on J. Let H be the system of all increasing homeomorphisms of J onto J and let $\mathcal{A}$ be the system of all functions $h \in H$ such that $f \circ h \in C_{ap}$ for each $f \in C_{ap}$. For each system $S \subset D$ let $\mathrm{Tr}\, S$ be the system of all functions $h \in H$ such that $f \circ h \in D$ for each $f \in S$.

The system $\mathcal{A}$ has been characterized in [1]. We shall see in 3.5 that $\mathcal{A} = \mathrm{Tr}\, \mathrm{b}\, C_{ap}$. Thus $\mathrm{Tr}\, S \subset \mathcal{A}$ for each S with $\mathrm{b}\, C_{ap} \subset S \subset D$. Theorem 3.6 describes $\mathrm{Tr}\, L$. It is easy to prove that we have even $f \circ h \in L$ for each $f \in L$ and each $h \in \mathrm{Tr}\, L$ (see 3.7). If $h \in H$ and if both functions h and h^{-1} satisfy the Lipschitz condition, then (9) is obviously fulfilled so that $h \in \mathrm{Tr}\, L$. In 3.8 we show that a function $h \in H$ belongs to $\mathrm{Tr}\, L$ if and only if $w \circ h \in W$ for each $w \in W$. In this way we obtain an improvement of Theorem 4 in [2]. Theorem 3.9 gives a characterization (which is simpler than the characterization found in [5]) of the system $\mathrm{Tr}\, D$.

We shall see in 4.7 that the obvious inclusions $\mathrm{Tr}\, D \subset \mathrm{Tr}\, L \subset \mathrm{Tr}\, \mathrm{b}\, C_{ap}$ $(= \mathcal{A})$ are proper. If, however, h is a convex or concave element of $\mathcal{A}$, then $h \in \mathrm{Tr}\, D$. This is proved in 3.12 (with the help of 3.11).

First we introduce an auxiliary system.

<u>3.1</u>. Let $\mathcal{B}$ be the system of all functions $g \in H \cap AC$ with the following property: If S is a measurable subset of J and if $x \in J$ is a point of dispersion for S, then $g(x)$ is a point of dispersion for $g(S)$.

Remark. It is easy to see that $g(S)$ is measurable whenever S is measurable, $S \subset J$ and $g \in AC$.

The assertions 3.2 and 3.3 follow easily from Theorems 1 and 2 and Lemma 6 in [1].

3.2. We have $h \in \mathcal{A}$ if and only if $h^{-1} \in \mathcal{B}$.

3.3. Let $h \in \mathcal{A}$, $a \in J$. Then there is a $\delta \in (0,1)$ such that $|h(x)-h(a)|/|x-a|^{\delta} \to 0$ $(x \to a,\ x \in J)$.

3.4. $\mathcal{A} \subset AC$.

Proof. Let $h \in \mathcal{A}$, $S \subset J$, $|S| = 0$. There is a G_δ set T such that $S \subset T \subset J$ and that $|T| = 0$. Suppose that $|h(T)| > 0$. The set $h(T)$ is obviously measurable. Let X be a point of density of $h(T)$. It follows easily from 3.2 that $h^{-1}(x)$ is a point of density of T; this, however, is impossible. Thus $|h(T)| = 0$, $|h(S)| = 0$, $h \in AC$.

3.5. $\mathcal{A} = \operatorname{Tr} b\,C_{ap}$.

Proof. If $h \in \mathcal{A}$ and $\alpha \in b\,C_{ap}$, then $\alpha \circ h \in b\,C_{ap} \subset D$ so that $h \in \operatorname{Tr} b\,C_{ap}$. Now let $h \in \operatorname{Tr} b\,C_{ap}$ and $\alpha \in C_{ap}$. Let $a, b \in R$, $\beta = (\alpha \wedge a) \vee b$. Then $\beta \in b\,C_{ap}$, $((\alpha \circ h) \wedge a) \vee b = \beta \circ h \in D$. By 1.7 we have $\alpha \circ h \in C_{ap}$. Thus $h \in \mathcal{A}$.

3.6. Let $h \in H$, $g = h^{-1}$. Then $h \in \operatorname{Tr} L$ if and only if

(9) $\limsup (g(x)-g(a))^{-1} \int_a^x \sup(Ug,t,x)\,dt < \infty \qquad (x \to a,\ x \in J)$

for each $a \in J$.

Proof. Suppose that (9) holds for each $a \in J$. It is easy to see that there is a finite set $S \subset J$ such that g fulfills the Lipschitz condition on each closed interval contained in $J \setminus S$. This shows that $g \in AC$ so that g is an indefinite integral of Ug. Let $f \in L$. It follows from 2.5 that $f \cdot Ug$ is L-integrable on J. Let $Q(x) = \int_0^x f \cdot Ug$ $(x \in J)$. By 2.5 we have

$$(g(x)-g(a))^{-1}(Q(x)-Q(a)) \to f(a) \qquad (x \to a,\ x \in J)$$

for each $a \in J$. Hence $(Q \circ h)' = f \circ h$ on J, $h \in \operatorname{Tr} L$.

Now suppose that, e.g., $\limsup (g(x))^{-1} \int_0^x \sup(Ug,t,x)\,dt = \infty$ $(x \to 0+)$. There are $c_n \in (0,1)$ such that $c_n \to 0$ and that $ng(c_n) < T_n$, where $T_n = \int_0^{c_n} \sup(Ug,t,c_n)\,dt$. There are $b_n \in (c_n, 2c_n) \cap (0,1)$ such that $ng(b_n) < T_n$ and that $g'(b_n)$ exists. Then $ng(b_n) < \int_0^{b_n} \sup(Ug,t,b_n)\,dt$ $(n = 1,2,\ldots)$. It follows easily from 2.7 that there is a subsequence $\langle a_n \rangle$ of $\langle b_n \rangle$ and nonnegative continuous functions f_n such that $2a_{n+1} < a_n$, $f_n = 0$ on $[0, a_{n+1}] \cup [a_n, 1]$, $\int_0^x f_n \leq x$ for $x \in J$ and that

$\int_0^{g(a_n)} f_n \circ h > ng(a_n)$ $(n = 1,2,\ldots)$. Set $A_n = g(a_n)$, $f = \Sigma\, n^{-1} f_n$. Let $a_n < x \leq a_{n-1}$. Then $\int_0^x f = (n-1)^{-1} \int_0^x f_{n-1} + \Sigma_{k=n}^{\infty} k^{-1} \int_0^{a_k} f_k \leq (n-1)^{-1} x + n^{-1} \Sigma_{k=n}^{\infty} a_k < (n-1)^{-1} x + n^{-1} 2a_n < (n-1)^{-1} 3x$. This shows that $f \in D$. It is easy to see that $f \in L$. However, $\int_0^{A_n} f \circ h \geq n^{-1} \int_0^{A_n} f_n \circ h > A_n$ so that $f \circ h \notin D$, $h \notin \mathrm{Tr}\, L$.

Remark. It follows from 3.6 that the set $\{x \in J;\ Lh(x) = 0\}$ is finite for each $h \in \mathrm{Tr}\, L$.

3.7. Let $h \in \mathrm{Tr}\, L$ and $f \in L$. Then $f \circ h \in L$.

Proof. Let $a \in R$. By 1.2, $f \vee a \in L$. Thus $f \circ h \in D$, $(f \circ h) \vee a \in D$. By 1.5 we have $f \circ h \in L$.

3.8. Let $h \in H$. Then $h \in \mathrm{Tr}\, L$ if and only if $w \circ h \in W$ for each $w \in W$.

Proof. If $h \in \mathrm{Tr}\, L$ and $w \in W$, then $w, w^2 \in L$ (see 1.9), $w \circ h$, $(w \circ h)^2 \in D$, $w \circ h \in W$.

Now suppose that $w \circ h \in W$ for each $w \in W$. Let $f \in L$. By 1.11 there are $v, w \in W$ such that $f = vw$. Thus $f \circ h = (v \circ h)(w \circ h) \in L$ (see 1.9), $h \in \mathrm{Tr}\, L$.

3.9. Let $h \in H$, $g = h^{-1}$. Let γ be a function such that $Lg \leq \gamma \leq Ug$ on J. Then the following three conditions are equivalent to each other:

i) There is a function φ such that $g(a) = \int_0^a \varphi$ and that

$$\limsup (g(x) - g(a))^{-1} \int_a^x \mathrm{var}(\varphi, t, x)\,dt < \infty \qquad (x \to a,\ x \in J)$$

for each $a \in J$;

ii) $h \in \mathrm{Tr}\, D$;

iii) the condition

$$\limsup (g(x) - g(a))^{-1} \int_a^x \mathrm{var}(\gamma, t, x)\,dt < \infty \qquad (x \to a,\ x \in J) \tag{10}$$

is fulfilled for each $a \in J$.

Proof. Suppose that i) holds. Let $f \in D$. It follows from 2.6 with $\alpha = \varphi$ that $f\varphi$ is P-integrable on J. Set $Q(x) = \int_0^x f\varphi$ $(x \in J)$. By 2.6 (with $\lambda = 1$) we have $(g(x) - g(a))^{-1} (Q(x) - Q(a)) \to f(a)$ $(x \to a,\ x \in J)$ for each $a \in J$. Hence $(Q \circ h)' = f \circ h$ on J, $h \in \mathrm{Tr}\, D$.

Now let $h \in \mathrm{Tr}\, D$. Suppose that, e.g.,

$$\limsup (g(x))^{-1} \int_0^x \mathrm{var}(\gamma, t, x)\,dt = \infty \qquad (x \to 0+).$$

There are $c_n \in (0,1)$ such that $c_n \to 0$ and that $ng(c_n) < V_n$, where $V_n = \int_0^{c_n} \mathrm{var}(\gamma, t, c_n)\,dt$. There are $b_n \in (c_n, 2c_n) \cap (0,1)$ such that

$ng(b_n) < V_n$ and that $g'(b_n)$ exists. Then $ng(b_n) < \int_0^{b_n} \text{var}(\gamma, t, b_n)dt$ $(n = 1,2,\ldots)$. It follows easily from 2.12 that there is a subsequence $\langle a_n \rangle$ of $\langle b_n \rangle$ and functions f_n continuous on J such that $a_{n+1} < a_n$, $f_n = 0$ on $[0, a_{n+1}] \cup [a_n, 1]$, $\int_J f_n = 0$, $|\int_0^x f_n| \leq x$ for each $x \in J$ and that

$$(11) \qquad \int_0^{g(a_n)} f_n \circ h > ng(a_n) \qquad (n = 1,2,\ldots).$$

Set $f = \Sigma n^{-1} f_n$, $F = 0$ on $\{0\} \cup (a_1, 1]$, $F(x) = n^{-1} \int_{a_{n+1}}^x f_n$ for $x \in (a_{n+1}, a_n]$. It is easy to see that $F' = f$ on J. By assumption there is a function G such that $G' = f \circ h$ on J. For $n = 1,2,\ldots$ set $q_n = (G(B) - G(A))/B$, where $A = g(a_{n+1})$, $B = g(a_n)$. Since $G'(0) = 0$, we have $q_n = (G(B)/B) - (A/B)(G(A)/A) \to 0$. However, by (11), $G(B) - G(A) = n^{-1} \int_A^B f_n \circ h > B$ whence $q_n > 1$ for each n. This contradiction proves iii).

Suppose, finally, that iii) holds. It is easy to see that there is a finite set $S \subset J$ such that $\text{var}(\gamma, K) < \infty$ for each closed interval $K \subset J \setminus S$. This shows that $g \in AC$. Thus (i) holds with $\varphi = \gamma$.

Remark 1. Suppose that $h \in H$, $h' > 0$ on J and that $\text{var}(h', J) < \infty$. Then $\text{var}(g', J) < \infty$ and (10) holds for each $a \in J$. Thus $h \in \text{Tr}\, D$. The remark in 3.6 and the example in 3.13 show that the requirement $h' > 0$ is essential.

Remark 2. Let $h \in \text{Tr}\, D$. It follows easily from 3.9 (see the proof of the implication iii) $\to$ i)) that there is a finite set $S \subset J$ such that for each $a \in (0,1) \setminus S$ the unilateral derivatives $h'^{+}(a)$, $h'^{-}(a)$ exist and that there is a countable set $T \subset J$ such that for each $a \in (0,1) \setminus T$ the derivative $h'(a)$ exists; we must, of course, admit also infinite derivatives. The next example shows that h' may be infinite on an uncountable set. (See also [5], p.195.)

<u>3.10</u>. Example of an $h \in \text{Tr}\, D$ such that $h' = \infty$ on a perfect set.

Let C be the Cantor set. Let G be a function such that $G(0) = 0$ and that $G'(x) = \text{dist}(x, C)$ for each $x \in J$. Then $|G''| = 1$ a.e. whence $\text{var}(G', t, x) = |x - t|$ for all $x, t \in J$. Let $a, b \in J$, $a < b$. Then $\int_a^b \text{var}(G', t, b)dt = \int_a^b (t-b)dt = (b-a)^2/2$. Set $q = 1/3$. There is an integer $n \geq 2$ such that $q^{n-1} < b - a \leq q^{n-2}$. Define $t_k = kq^n$ $(k = 0,\ldots,3^n)$. There are j, k such that $t_{j-1} \leq a < t_j$, $t_k < b \leq t_{k+1}$. It is easy to see that $k \geq j + 2$ and that at least one of the intervals (t_j, t_{j+1}), (t_{j+1}, t_{j+2}) is contained in $J \setminus C$.

Let, e.g., $I \cap C = \emptyset$ with $I = (t_j, t_{j+1})$. For $x \in I$ set $\mu(x) = \min(x-t_j, t_{j+1}-x)$. Then $I \subset (a,b)$, $|I| = q^n \geq (b-a)/9$, $G' \geq \mu$ on I and $G(b)-G(a) \geq \int_I \mu = |I|^2/4 \geq (b-a)^2/324$. Hence $(G(b)-G(a))^{-1}\int_a^b \operatorname{var}(G',t,b)dt \leq 162$. Now set $g = 28G$, $h = g^{-1}$. Since $G(1) = \Sigma_{n=1}^{\infty} 2^{n-1}(q^n)^2/4 = 1/28$, we have $h \in H$. By 3.9, $h \in \operatorname{Tr} D$. Obviously $h' = \infty$ on $g(C)$.

3.11. Let $g \in \mathcal{B}$ and let g be convex. Then

$$\limsup xg'^{+}(x)/g(x) < \infty \qquad (x \to 0+).$$

Proof. Suppose that the assertion is false. Then there are $a_n \in (0,1)$ such that $2a_n < a_{n-1}$ and $a_n g'^{+}(a_n) > ng(a_n)$ for $n = 1,2,\ldots$. Set $b_n = (1+n^{-1})a_n$, $A_n = g(a_n)$, $B_n = g(b_n)$, $S = \bigcup(a_n,b_n)$. It is easy to see that $|S \cap (0,x)|/x \to 0$ $(x \to 0+)$. However, $B_n - A_n \geq (b_n-a_n)g'^{+}(a_n) > ng(a_n)(b_n-a_n)/a_n = A_n$ so that $|g(S) \cap (0,B_n)| > B_n - A_n > B_n/2$. Thus 0 is a point of dispersion for S, but not for $g(S)$ which is a contradiction.

3.12. (Cf. [5], Theorem 5.) Let $h \in H$ and let h be concave. Then the following three conditions are equivalent to each other: (i) $h \in \operatorname{Tr} D$; (ii) $h \in \mathcal{A}$; (iii) $\limsup h(x)/(xh'^{+}(x)) < \infty$ $(x \to 0+)$.

Proof. The implication (i) $\to$ (ii) follows from 3.5; the implication (ii) $\to$ (iii) follows from 3.2 and 3.11. Now suppose that (iii) holds. Set $g = h^{-1}$, $\gamma = Ug$ $(= g'^{+}$ on $[0,1])$. We prove first that (10) holds for $a = 0$. Let $0 < t < x < 1$. Since $\operatorname{var}(\gamma,t,x) \leq \gamma(x)$, we have $(g(x))^{-1}\int_0^x \operatorname{var}(\gamma,t,x)dt \leq x\gamma(x)/g(x)$ and (10) follows from (iii). The reader easily verifies that (10) holds for each $a \in (0,1]$. Now (i) follows from 3.9.

Remark 1. It follows at once from 3.3 that there are concave functions $h \in H\backslash\mathcal{A}$; by 3.12, such an h does not fulfill (iii). It is, however, easy to construct a concave function $h \in H$ violating (iii) such that $h(x) \leq x^{1/2}$ $(x \in J)$.

Remark 2. Condition (iii) is certainly fulfilled (for h in H and concave), if $h'^{+}(0) < \infty$. In particular, each differentiable concave element of H is in $\operatorname{Tr} D$. Similarly, $h \in \operatorname{Tr} D$ for each differentiable convex function $h \in H$. If, however, $h \in H$ and if h is the difference of two differentiable convex functions, we need not have even $h \in \mathcal{A}$ as the following example shows.

3.13. Example of an $h \in H\backslash\mathcal{A}$ such that h'' is continuous on J and $h' > 0$ on $(0,1]$.

Let $a_n = 3^{-n}$ $(n = 0,1,\ldots)$. Set $\varphi(x) = x - \sin x$,

$\lambda_n(x) = 2\pi a_n^{-1}(x-a_n)$, $\alpha_n = 10^{-n}$, $\beta_n = 9\alpha_n$. Note that $\varphi'(0) = \varphi'(2\pi) = \varphi''(0) = \varphi''(2\pi) = 0$ and that φ increases. Let $\psi(0) = 0$, $\psi = \alpha_n + \beta_n(2\pi)^{-1}\varphi\circ\lambda_n$ on $(a_n, 2a_n]$, $\psi = \alpha_{n-1}$ on $(2a_n, 3a_n]$. Obviously $\psi(a_n+) = \alpha_n$, $\psi(2a_n) = \alpha_n + \beta_n = \alpha_{n-1}$ so that ψ is continuous and nondecreasing on J. We have $\psi' = \beta_n 3^n \varphi'\circ\lambda_n$, $\psi'' = 2\pi\beta_n 9^n \varphi''\circ\lambda_n = 18\pi(9/10)^n \sin\circ\lambda_n$ on $(a_n, 2a_n)$. We see that the functions ψ' and ψ'' are continuous on J as well. Set $h(x) = \frac{1}{2}(x^3 + \psi(x))$ $(x \in J)$. Then $h \in H$, $h' > 0$ on $(0,1]$ and h'' is continuous on J. Set $A_n = h(2a_n)$, $B_n = h(3a_n)$ $(n = 1,2,\ldots)$, $S = \bigcup(A_n, B_n)$. Obviously $B_n - A_n = \frac{19}{2}27^{-n}$. If $A_n < x \leq A_{n-1}$, then $|S \cap (0,x)| \leq \sum_{k=n}^{\infty}(B_k - A_k) = \frac{19}{52}27^{1-n}$. Since $A_n > \psi(2a_n)/2 = 10^{1-n}/2$, we have $|S \cap (0,x)|/x \to 0$ $(x \to 0+)$. However, $h^{-1}(S) = \bigcup(2a_n, 3a_n)$. This shows that $h^{-1} \notin \mathcal{B}$. By 3.2 we have $h \in H\setminus\mathcal{Q}$.

4. MULTIPLICATION. For each system $S \subset D$ let $\mathrm{Mu}\,S$ be the family of all functions α such that $f\alpha \in D$ for each $f \in S$. There is a close connection between $\mathrm{Mu}\,S$ and $\mathrm{Tr}\,S$. To see this, choose functions $f \in D$ and $h \in H$ such that $0 < h' < \infty$ on J. Let $g = h^{-1}$. It follows from the chain rule that $fg' \in D$ if and only if $f\circ h \in D$. This shows that, for any $S \subset D$, we have $h \in \mathrm{Tr}\,S$ if and only if $g' \in \mathrm{Mu}\,S$. This observation helps us to describe $\mathrm{Mu}\,D$ (see 4.5). We need first the auxiliary assertion 4.1 from which we obtain easily in 4.2 the result $\mathrm{Mu}\,L = bD$. It is also true that $\mathrm{Mu}\,bD = L$. This, however, will be proved elsewhere together with the description of systems $\mathrm{Mu}\,S$ for some other families $S \subset D$.

<u>4.1</u>. Let α be a function on J such that $\limsup \alpha(x) = \infty$ $(x \to 0+)$. Then there is an $f \in D$ such that $f(0) = 0$, f is continuous and nonnegative on $(0,1]$ (in particular, $f \in L$) and $f\alpha \notin D$.

Proof. If α is not a derivative on $(0,1]$, then there is an $a \in (0,1)$ such that α is not a derivative on $(a,1]$. Then the function f such that $f(x) = x$ on $[0,a)$ and $f = a$ on $[a,1]$ fulfills our requirements.

Now suppose that α is a derivative on $(0,1]$. There are $a_n \in (0,1)$ such that $2a_n < a_{n-1}$ and $\alpha(a_n) > n$. There are $b_n \in (a_n, 2a_n)$ such that $\int_{a_n}^{b_n} \alpha > n(b_n - a_n)$ $(n = 1,2,\ldots)$. There is a function f continuous and nonnegative on $(0,1]$ such that $f = a_n/(n(b_n - a_n))$ on (a_n, b_n) and that $\int_{a_n}^{a_{n-1}} f < 2a_n/n$. If

$a_n < x \le a_{n-1}$, then $x^{-1}\int_0^x f \le a_n^{-1}\int_0^{a_{n-1}} f < 4/n$. Set $f(0)=0$. Then $f \in D$. Suppose that there is a Q such that $Q' = f\alpha$ on J. We may suppose that $Q(0)=0$. Obviously $Q'(0)=0$ so that $(Q(b_n)-Q(a_n))/b_n = (Q(b_n)/b_n)-(a_n/b_n)Q(a_n)/a_n \to 0$. However, $Q(b_n)-Q(a_n) = (a_n/(n(b_n-a_n)))\int_{a_n}^{b_n}\alpha > a_n > b_n/2$ for each n which is a contradiction.

4.2. $\mathrm{Mu}\,L = bD$.

Proof. Let $f \in L$, $\alpha \in D$, $|\alpha| \le 1$, $c \in J$. Then $|(x-c)^{-1}\int_c^x (f-f(c))\cdot\alpha| \le (x-c)^{-1}\int_c^x |f-f(c)| \to 0$, $(x-c)^{-1}\int_c^x f(c)\cdot\alpha \to f(c)\alpha(c)$ $(x \to c,\ x \in J)$. Thus $f\alpha \in D$, $\alpha \in \mathrm{Mu}\,L$.

Now let $\alpha \in \mathrm{Mu}\,L$. It is obvious that $\alpha \in D$ and it follows easily from 4.1 that α is bounded.

4.3. $\mathrm{Mu}\,D \subset bC_{ap}$. (See [4], Theorems 4 and 8.)

Proof. Obviously $\mathrm{Mu}\,D \subset W \cap \mathrm{Mu}\,L$. Now we apply 1.4, 1.3 and 4.2.

4.4. Let ψ be a finite nonincreasing function on $(0,1)$. Let $A = \limsup(\psi(x)-\psi(2x))$, $B = \limsup x^{-1}\int_0^x(\psi(t)-\psi(x))dt$ $(x \to 0+)$. Then $A \le 2B$, $B \le 2A$.

Proof. If $A < A_1 < \infty$, then there is a $\delta \in (0,1)$ such that $\psi(x/2)-\psi(x) < A_1$, whenever $0 < x < \delta$. Choose such an x. Obviously $\psi(x/2^n)-\psi(x) < nA_1$ for $n = 1,2,\ldots$ so that $\int_0^x(\psi(t)-\psi(x))dt < \Sigma_{n=1}^{\infty} nA_1 x/2^n = 2A_1 x$. Thus $B \le 2A_1$, $B \le 2A$.

If $B < B_1 < \infty$, then there is a $\delta \in (0,1)$ such that $\int_0^{2x}(\psi(t)-\psi(2x))dt < 2B_1 x$, whenever $0 < x < \delta$. Choose such an x. Then $2B_1 x > \int_0^x(\psi(x)-\psi(2x))dt = x(\psi(x)-\psi(2x))$ so that $A \le 2B_1$, $A \le 2B$.

4.5. Let $\alpha \in D$. Then $\alpha \in \mathrm{Mu}\,D$ if and only if

(12) $\limsup \mathrm{var}(\alpha,(a+x)/2,x) < \infty$ $(x \to a,\ x \in J)$ for each $a \in J$.

Proof. If (12) holds, then, by 4.4, $\limsup(x-a)^{-1}\int_a^x \mathrm{var}(\alpha,t,x)dt < \infty$ for each $a \in J$. Let $f \in D$, $a \in J$. It follows from 2.6 with $\varphi = 1$, $\lambda = \alpha(a)$ that $f\alpha$ is P-integrable and that $(x-a)^{-1}\int_a^x f\alpha \to (f\alpha)(a)$ $(x \to a,\ x \in J)$. Thus $f\alpha \in D$, $\alpha \in \mathrm{Mu}\,D$.

Now let $\alpha \in \mathrm{Mu}\,D$. It follows from 4.3 that there is a $c \in R$ such that $\alpha + c > 0$ on J. Let $\gamma = \alpha + c$, $g' = \gamma/\int_J \gamma$, $g(0) = 0$. Since $g \in H$ and $g' \in \mathrm{Mu}\,D$, we have $g^{-1} \in \mathrm{Tr}\,D$ whence, by 3.9,

$$\limsup(g(x)-g(a))^{-1}\int_a^x \mathrm{var}(g',t,x)dt < \infty \text{ for each } a \in J.$$

Therefore $\limsup(x-a)^{-1}\int_a^x \mathrm{var}(\alpha,t,x)dt < \infty$ $(x \to a,\ x \in J)$ for each $a \in J$ and (12) follows from 4.4.

4.6. Example of an $h \in H \backslash \mathrm{Tr}\, D$ such that $\frac{1}{2} < h' < 2$ on J. (Cf. [2], Example 2.)

Let $\alpha \in D \backslash C_{ap}$, $2^{-1/2} < \alpha < 2^{1/2}$. (We may choose, e.g., $\alpha(0) = 1$, $\alpha(x) = 1 + \frac{1}{4} \sin x^{-1}$ for $x \in (0,1]$.) Let $g' = \alpha / \int_J \alpha$, $g(0) = 0$, $h = g^{-1}$. From $a^{\alpha} \in D$ it would follow (see 1.4 and 1.3) that $\alpha \in C_{ap}$; thus $\alpha g' \notin D$, $\alpha \circ h \notin D$, $h \in H \backslash \mathrm{Tr}\, D$. Obviously $\frac{1}{2} < g' < 2$, $\frac{1}{2} < h' < 2$ on J.

4.7. The inclusions $\mathrm{Tr}\, D \subset \mathrm{Tr}\, L \subset \mathrm{Tr}\, b\, C_{ap}$ $(= \mathcal{A})$ are proper.

Proof. It is easy to see that there is a nonnegative function $\varphi \in D$ such that $\varphi^2 \in D$, $\varphi^3 \notin D$. (Such a φ may be continuous on $(0,1]$.) Let $\psi = \varphi + 1$, $g' = \psi / \int_J \psi$, $g(0) = 0$, $h = g^{-1}$. Choose an $\alpha \in b\, C_{ap}$. Since $\alpha, \psi \in W$, we have (see 1.9) $\alpha g' \in L$ so that $\alpha \circ h \in D$, $h \in \mathrm{Tr}\, b\, C_{ap}$. However, $\varphi^2 g' \notin D$ so that $\varphi^2 \circ h \notin D$. Thus (since $\varphi^2 \in L$) $h \notin \mathrm{Tr}\, L$. This shows that the second inclusion is proper. It follows from 4.6 that the first inclusion is proper.

BIBLIOGRAPHY

1. A. M. Bruckner, "Density-preserving homeomorphisms and a theorem of Maximoff," Quart. J. Math. Oxford (2), 21 (1970), 337-347.

2. ________, "On transformations of derivatives," Proc. Amer. Math. Soc. 48 (1975), 101-107.

3. ________, Differentiation of real functions, Lecture Notes in Mathematics, 659, Springer-Verlag, Berlin Heidelberg New York, 1978.

4. R. J. Fleissner, "Distant bounded variation and products of derivatives," Fund. Math. 94 (1977), 1-11.

5. M. Laczkovich and G. Petruska, "On the transformers of derivatives," Fund. Math. (1978), 179-199.

6. S. Saks, Theory of the integral, Monografie Matematyczne 7, Warszawa-Lwów, 1937.

DEPARTMENT OF MATHEMATICS
MICHIGAN STATE UNIVERSITY
EAST LANSING, MICHIGAN 48824

Contemporary Mathematics
Volume 42, 1985

A LUSIN TYPE APPROXIMATION OF SOBOLEV FUNCTIONS BY SMOOTH FUNCTIONS

J.H. MICHAEL and WILLIAM P. ZIEMER[1]

ABSTRACT. In this paper it is shown that Sobolev functions can be approximated by smooth functions both in norm and capacity.

1 INTRODUCTION

In [11] Michael proved the following.

Let f be a measurable real-valued function on R^n, which vanishes outside a compact set and whose Lebesgue area $A(f,E)$ is finite on compact sets E. Let $\varepsilon > 0$ and Q be a closed cube. Then there exists a Lipschitz function g on R^n with compact support and such that

$$m\{x;\ f(x) \neq g(x)\} < \varepsilon$$

and

$$|A(f,Q) - A(g,Q)| < \varepsilon .$$

This proof answered in the affirmative a conjecture made by Casper Goffman. Subsequently, it was shown by Goffman in [6] that the approximating function g in the above result could be taken to be C^1. In [12] and [13] Michael proved similar theorems for functionals different from the area functional.

Another Goffman conjecture of a similar nature was settled in the affirmative by Fon-Che-Liu in [7] when he proved the following.

If G is a strongly Lipschitz domain in R^n, $1 \leq p < \infty$, $f \in W^{\ell,p}(G)$ and $\varepsilon > 0$, then there exists a C^ℓ function g on G, such that

(i) $$m\{x;\ x \in G \text{ and } f(x) \neq g(x)\} < \varepsilon \text{ and}$$

(ii) $$|f-g|_{\ell,p} < \varepsilon .$$

1980 Mathematics Subject Classification. 26B35, 41A30, 46E35.

[1]Research supported in part by a grant from the National Science Foundation.

This was generalized by Michael in an unpublished paper to the case where G is an arbitrary open subset of R^n. Michael's paper also gave an independent proof of Liu's theorem.

In the present paper Liu's theorem is generalized in such a way that the approximation (i) is made with respect to capacity instead of measure. However, since a Sobolev function may be undefined on a set of measure zero and this set could have positive capacity, one has to begin by finding a special representative for the Sobolev function. We therefore show in Theorem 3.4 that:

if $1 \leq p < \infty$, ℓ is a positive integer, $\ell p < n$, Ω is an open set of R^n and $f \in W^{\ell,p}(\Omega)$, then there exists a function g, with $g(x) = f(x)$ for almost all $x \in \Omega$ and such that g is approximately continuous at every point of Ω, except for a set E with $R_{\ell,p}(E) = 0$. Here, $R_{\ell,p}$ denotes Riesz capacity. This leads to Theorem 3.8, which generalizes a result of Federer and Ziemer in [4]. Our main result, Theorem 4.6, can now be described as follows.

Let $1 \leq p < \infty$, let ℓ, m be positive integers with $1 \leq m \leq \ell$ and $(\ell-m)p < n$ and let Ω be an open set of R^n. Let $f \in W^{\ell,p}(\Omega)$ and be approximately continuous at every point of Ω except for a set E with $R_{\ell-m,p}(E) = 0$. Let $\varepsilon > 0$. Then there exists a C^m function g on Ω such that

(a) the set $F = \{x;\ x \in \Omega \text{ and } f(x) \neq g(x)\}$ has $R_{\ell-m,p}(F) < \varepsilon$ and

(b) $|f-g|_{m,p} < \varepsilon$.

Since we regard $R_{0,p}$ as being Lebesgue measure, Liu's theorem is the special case of our theorem with $m = \ell$.

Another theorem of some interest is 3.10. In this we let p, ℓ, m and f be as in the main theorem but let the domain Ω be the whole of R^n. We derive a Taylor expansion of order m for f about every point of R^n, except for a set E, with $R_{\ell-m,p}(E) = 0$. Since this is valid when $1 \leq p < \infty$ it covers cases which are not included in the Taylor expansion of Meyers in [9].

2 PRELIMINARIES

The paper will be concerned with subsets of R^n, where $n \geq 1$ and with $W^{\ell,p}$ and L^p spaces, where it will always be assumed that $1 \leq p < \infty$. In proving the various theorems, Riesz capacity appears to be more naturally applicable than Bessel capacity, so we use the former. We follow Stein [14,p.117] and define, for $0 < k < n$ and $f \in L^p(R^n)$, the Riesz potential $I_k(f)$ of f by putting

$$(I_k(f))(x) = \frac{1}{\gamma(k)} \int_{R^n} |x-y|^{k-n} f(y)dy \ . \tag{1}$$

$\gamma(k)$ is a positive constant. Its precise value is given in [14], but since we will be dealing only with sets of small capacity, the actual value of $\gamma(k)$ is not important here.

(A) when $f \geq 0$, $I_k(f)$ is lower semi-continuous on R^n . In the usual way, we define the Riesz capacity $R_{k,p}(E)$ of a subset E of R^n by

$$R_{k,p}(E) = \inf(|f|_p)^p \tag{2}$$

where the infimum is taken over all non-negative $f \in L^p(R^n)$ such that $(I_k f)(x) \geq 1$ when $x \in E$. It is well known and follows from (A) that

(B) if E is a subset of R^n with $R_{k,p}(E) < \infty$ and if $\varepsilon > 0$, then there exists an open set U , such that $U \supset E$ and

$$R_{k,p}(U) < R_{k,p}(E) + \varepsilon \ .$$

For each subset E of R^n we regard $R_{0,p}(E)$ as being the ordinary Lebesgue measure of E . Clearly (B) holds in this case also.

The following further properties of Riesz capacity are also well known (In each statement $k \geq 0$ and $kp < n$) .

(C) Every bounded subset E of R^n has

$$R_{k,p}(E) < \infty \ .$$

(D) If B is the unit ball in R^n , then

$$R_{k,p}(E) > 0 \ .$$

(E) There exists a positive constant C , depending only on n , k and p and such that

$$R_{k,p}(B) = C\rho^{n-kp} \ ,$$

for every closed ball B in R^n with radius ρ .

For $a > 0$, let H^α denote α-dimensional Hausdorff measure.

(F) There exists a positive constant C , depending only on n , k and p and such that

$$R_{k,p}(E) \leq CH^{n-kp}(E)$$

for every subset E of R^n, for which $H^{n-kp}(E) < \infty$.

3 A FIRST APPROXIMATION TO SOBOLEV FUNCTIONS

In 3 we prove part of the main approximation theorem and some of the other results mentioned in 1.

3.1 THEOREM

Let B *be an open ball with centre* x' *and radius* ρ, *let* $f \in W^{1,p}(B)$ *and let* $0 < \delta < \rho$. *Then*

$$\rho^{-n} \int_{|y-x'|<\rho} f(y)dy - \delta^{-n} \int_{|y-x'|<\delta} f(y)dy$$
$$= -\frac{1}{n} \delta^{-n} \int_{|y-x'|<\delta} [Df(y)\cdot(y-x')]dy$$
$$+ \frac{1}{n} \rho^{-n} \int_{|y-x'|<\rho} [Df(y)\cdot(y-x')]dy$$
$$- \frac{1}{n} \int_{\delta<|x-y'|<\rho} |y-x'|^{-n}[Df(y)\cdot(y-x')]dy . \quad (3)$$

PROOF: Define μ on R by

$$\mu(t) = \delta^{-n} - \rho^{-n} \quad \text{when } t \le \delta ,$$
$$= t^{-n} - \rho^{-n} \quad \text{when } \delta < t \le \rho ,$$
$$= 0 \quad \text{when } t > \rho .$$

Then

$$\int_{R^n} f(y)[\frac{\partial}{\partial y_i}\{\mu(|y-x'|)(y_i-x_i')\}]dy$$
$$= - \int_{R^n} \frac{\partial f}{\partial y_i}(y)\mu(|y-x'|)(y_i-x_i')dy . \quad (4)$$

By summing both sides of (4) with respect to i and computing

$$\sum_{i=1}^{n} \frac{\partial}{\partial y_i} \{\mu(|y-x'|)(y_i-x_i')\}$$

one can obtain the equation (3).

3.2 LEMMA

Let k *be a positive real number such that* $kp < n$ *and let* $f \in W^{1,p}(R^n)$. *Then*

$$\int_{R^n} |y-x|^{k-n}|f(y)|dy \le \frac{1}{k}\int_{R^n} |y-x|^{k-n+1}|Df(y)|dy . \tag{5}$$

for all $x \in R^n$.

PROOF: (i) We suppose first of all that f vanishes outside a bounded set. Let $x \in R^n$ and for $r = 1,2,\ldots$ define a C^∞ function ϕ_r on R^n by

$$\phi_r(y) = [\tfrac{1}{r} + |y-x|^2]^{\frac{1}{2}(k-n)} .$$

Since $|f| \in W^{1,p}(R^n)$, it follows that

$$\int_{R^n} [\sum_{i=1}^{n} \frac{\partial}{\partial y_i}\{\phi_r(y)(y_i-x_i)\}]|f(y)|dy$$

$$= -\int_{R^n} \phi_r(y)[(y-x)\cdot D|f|(y)]dy ,$$

so that, since $|D|f|| = |Df|$,

$$\int_{R^n} [\sum_{i=1}^{n} \frac{\partial}{\partial y_i} \{\phi_r(y)(y_i-x_i)\}]|f(y)|dy$$

$$\le \int_{R^n} \phi_r(y)|y-x||Df(y)|dy . \tag{6}$$

By calculating the partial derivatives on the left-hand side of (6) one obtains

$$\int_{R^n} [\tfrac{1}{r} + |y-x|^2]^{\frac{1}{2}(k-n-2)} [k|y-x|^2 + \tfrac{n}{r}]dy$$

$$\le \int_{R^n} [\tfrac{1}{r} + |y-x|^2]^{\frac{1}{2}(k-n)} |y-x||Df(y)|dy .$$

The inequality (5) now follows, in this case, when $r \to \infty$.

(ii) The general case. Let η be a C^∞ function on R , such that $0 \le \eta \le 1$, $\eta(t) = 1$ when $t \le 1$ and $\eta(t) = 0$ when $t \ge 2$. Define

$$f_r(y) = f(y)\eta(r^{-1}|y|)$$

for $y \in R^n$. By applying (i) to f_r and then letting $r \to \infty$, one can verify (5) in the general case.

3.3 LEMMA

Let k *be a positive real number and* ℓ *a positive integer such that* $(k+\ell-1)p < n$. *Then there exists a constant* C *, depending only on* n *,* k

and ℓ *and such that*

$$\int_{R^n} |y-x|^{k-n}|f(y)|dy \le C \sum_{|\alpha|=\ell} \int_{R^n} |y-x|^{k-n+\ell}|D^\alpha f(y)|dy$$

for all $x \in R^n$ *and all* $f \in W^{\ell,p}(R^n)$.

This follows from 3.2 by mathematical induction.

We are now in a position to prove the theorem on approximate continuity that was mentioned in the introduction.

3.4 THEOREM

Let ℓ *be a positive integer such that* $\ell p < n$, *let* Ω *be a non-empty open subset of* R^n *and let* $f \in W^{\ell,p}(\Omega)$. *Then there exists a subset* E *of* Ω , *such that*

$$R_{\ell,p}(E) = 0$$

and

$$\lim_{\delta\to 0+} \delta^{-n} \int_{|y-x|<\delta} f(y)dy \tag{7}$$

exists for all $x \in \Omega \sim E$.

PROOF: (i) We suppose first of all that $\Omega = R^n$. Define

$$g(y) = \sum_{|\alpha|=\ell} |D^\alpha f(y)| \tag{8}$$

for $y \in R^n$. Then $g \in L^p(R^n)$. Let E be the set of all those points x of R^n for which

$$(I_\ell g)(x) = \infty . \tag{9}$$

Then

$$R_{\ell,p}(E) = 0 .$$

Consider an $x \in R^n \sim E$. By 3.1,

$$\int_{|y-x|<1} f(y)dy - \delta^{-n} \int_{|y-x|<\delta} f(y)dy = -\frac{1}{n}\delta^{-n} \int_{|y-x|<\delta} [Df(y)\cdot(y-x)]dy$$

$$+ \frac{1}{n} \int_{|y-x|<1} [Df(y)\cdot(y-x)]dy - \frac{1}{n} \int_{\delta<|y-x|<1} |y-x|^{-n}[Df(y)\cdot(y-x)]dy . \tag{10}$$

When $\ell = 1$, it follows from (8) and (9) that

$$\int_{|y-x|<1} |y-x|^{1-n} |Df(y)| dy < \infty . \tag{11}$$

When $\ell > 1$, it follows from 3.3, with $k = 1$ and $\ell - 1$ substituted for ℓ, that

$$\int_{R^n} |y-x|^{1-n} \left|\frac{\partial f}{\partial y_i}(y)\right| dy$$

$$\leq C \sum_{|\alpha|=\ell-1} \int_{R^n} |y-x|^{\ell-n} \left|D^\alpha \frac{\partial f}{\partial y_i}(y)\right| dy ,$$

which, by (8) and (9), is finite. Thus (11) still holds when $\ell > 1$.

By (11)

$$\lim_{\delta \to 0+} \int_{\delta<|y-x|<1} |y-x|^{-n} [Df(y) \cdot (y-x)] dy \tag{12}$$

exists. It also follows from (11) that

$$\lim_{\delta \to 0+} \int_{|y-x|<\delta} |y-x|^{1-n} |Df(y)| dy = 0 ,$$

hence

$$\delta^{-n} \int_{|y-x|} [Df(y) \cdot (y-x)] dy \to 0 \tag{13}$$

as $\delta \to 0+$. It now follows from (10), (12) and (13) that the limit in (7) exists.

(ii) The general case. Let Ω be an open set of R^n. There exists an increasing sequence $\{\phi_r\}$ of non-negative C^∞ functions on R^n, with compact supports, with $\operatorname{spt} \phi_r \in \Omega$ for all r and such that the interiors of the sets

$$\{x \ ; \ x \in R^n \text{ and } \phi_r(x) = 1\}$$

tend to Ω as $r \to \infty$. Define

$$f_r(x) = \phi_r(x) \cdot f(x) \quad \text{when} \quad x \in \Omega$$

$$= 0 \quad \text{when} \quad x \notin \Omega .$$

By applying (i) to each of the functions f_r, one can easily prove the theorem in this case.

The following covering lemma is needed. Although it is not the same as the lemma in I 1.6 of [14], it is proved in I 1.7 of [14].

3.5 LEMMA

Let E *be a non-empty subset of* R^n *and suppose that* $\mathcal{B}$ *is a collection of closed balls such that:*

(i) for every $x \in E$ there exists a $B \in \mathcal{B}$ with centre x ;

(ii) there exists a constant K , such that

$$\operatorname{diam} B \le K$$

for all $B \in \mathcal{B}$ and

(iii) for every infinite sequence $\{B_r\}$ in $\mathcal{B}$, with $B_r \cap B_s = \emptyset$ when $r \ne s$, it is true that

$$\operatorname{diam} B_r \to 0$$

as $r \to \infty$.

For each $B \in \mathcal{B}$, let B^* be the closed ball with the same centre but with 5 times the radius.

Then there exists a sequence $\{B_r\}$ in $\mathcal{B}$, such that $B_r \cap B_s = \emptyset$ when $r \ne s$ and

$$E \in \bigcup_r B_r^* .$$

3.6 LEMMA

Let ℓ *be a non-negative integer such that* $\ell p < n$ *and* λ *be a real number such that* $\ell < \lambda < \frac{n}{p}$. *Let* $f \in W^{\ell,p}(R^n)$. *Then*

$$\delta^{\lambda-\ell-n} \int_{|y-x|<\delta} f(y)dy \to 0 \tag{14}$$

as $\delta \to 0+$, *for all* $x \in R^n$ *except for a set* E *with* $R_{\lambda,p}(E) = 0$.

PROOF: (i) Suppose $\ell = 0$. It follows from Bagby and Ziemer [1], Corollary 3.4 (i), that

$$\delta^{p\lambda-n} \int_{|y-x|<\delta} |f(y)|^p \, dy \to 0 \tag{15}$$

as $\delta \to 0+$, for all $x \in R^n$ except for a set E with $R_{\lambda,p}(E) = 0$. Consider an $x \in R^n \sim E$. Clearly, when $p = 1$, the inequality

$$\delta^{\lambda-n}\left|\int_{|y-x|<\delta} f(y)dy\right| \le A\left[\delta^{\lambda p-n}\int_{|y-x|<\delta} |f(y)|^p\,dy\right]^{\frac{1}{p}} \tag{16}$$

holds. When $p > 1$, the inequality (16) can be verified by using Holder's inequality (A depends only on n and p). (14) now follows from (15) and (16).

(ii) Now suppose $\ell > 0$. Let E be the set of all x for which

$$\int_{R^n} |y-x|^{\lambda-n}\left[\sum_{|\alpha|=\ell}\left|D^\alpha f(y)\right|\right]dy = \infty . \tag{17}$$

Then $R_{\lambda,p}(E) = 0$.

Consider an $x \in R^n \sim E$. When $\ell = 1$, it follows from (17) that

$$\int_{|y-x|<1} |y-x|^{\lambda-\ell+1-n}|Df(y)|dy < \infty . \tag{18}$$

when $\ell > 1$, we replace k by $\lambda - \ell + 1$ and ℓ by $\ell - 1$ in 3.3 and again derive (18) from (17). We now show that

$$\delta^{\lambda-\ell}\int_{\delta<|y-x|<1} |y-x|^{1-n}|Df(y)|dy \to 0 , \tag{19}$$

as $\delta \to 0+$. Let $\rho \in (0,1)$ be arbitrary. Clearly

$$\delta^{\lambda-\ell}\int_{\rho\le|y-x|<1} |y-x|^{1-n}|Df(y)|dy \to 0 \tag{20}$$

as $\delta \to 0+$. When $0 < \delta < \rho$, we have

$$\delta^{\lambda-\ell}\int_{\delta<|y-x|<\rho} |y-x|^{1-n}|Df(y)|dy \le \int_{|y-x|<\rho} |y-x|^{\lambda-\ell+1-n}|Df(y)|dy . \tag{21}$$

It follows from (18) that the right-hand side of (21) approaches zero as $\rho \to 0+$. (19) now follows from (20) and (21).

Since $\lambda - \ell + 1 - n < 0$, it follows from (18) that

$$\int_{|y-x|<1} |y-x|\,|Df(y)|dy < \infty \tag{22}$$

Since $\lambda - \ell - n < 0$, it follows that

$$\delta^{\lambda-\ell-n} \int_{|y-x|<\delta} |y-x||Df(y)|dy \le \int_{|y-x|<\delta} |y-x|^{\lambda-\ell+1-n}|Df(y)|dy ,$$

so that by (18),

$$\delta^{\lambda-\ell-n} \int_{|y-x|<\delta} |y-x||Df(y)|dy \to 0 \tag{23}$$

as $\delta \to 0+$. By putting $\rho = 1$ in (3), one can obtain (14) from (3), (19) and (23).

3.7 THEOREM

Let k, ℓ *be integers such that* $k \ge 1$, $0 \le \ell \le k$ *and* $kp < n$. *Let* $f \in W^{\ell,p}(R^n)$ *and for each* $x \in R^n$ *and* $\rho > 0$ *put*

$$f_{\rho,x} = \frac{1}{m(B_\rho)} \int_{|y-x|<\rho} f(y)dy ,$$

where B_ρ *denotes the open ball with centre* 0 *and radius* ρ. *Then*

$$\rho^{(k-\ell)p-n} \int_{|y-x|<\rho} |f(y)-f_{\rho,x}|^p \, dy \to 0 \tag{24}$$

as $\rho \to 0+$, *for all* $x \in R^n$ *except for a set* E *with* $R_{k,p}(E) = 0$.

PROOF: We proceed by induction on ℓ. Suppose to begin with that $\ell = 0$. It follows from Bagby and Ziemer [1], Corollary 3.4 (i) that

$$\rho^{kp-n} \int_{|y-x|<\rho} |f(y)|^p \, dy \to 0 \tag{25}$$

for all $x \in R^n$, except for a set E' with

$$R_{k,p}(E') = 0 . \tag{26}$$

We now have, for $x \in R^n \sim E'$,

$$\rho^{(k-\frac{n}{p})}\left[\int_{|y-x|<\rho} |f(y)-f_{p,x}|^p \, dy\right]^{\frac{1}{p}} \le \rho^{(k-\frac{n}{p})}\left[\int_{|y-x|<\rho} |f(y)|^p \, dy\right]^{\frac{1}{p}}$$

$$+ \, p^{(k-\frac{n}{p})} |f_{p,x}| \left[\int_{|y-x|<\rho} dy\right]^{\frac{1}{p}} . \tag{27}$$

But by 3.6

$$\rho^k f_{\rho,x} \to 0 \tag{28}$$

as $\rho \to 0+$, for all $x \in R^n$ except for a set E'' with $R_{k,p}(E'') = 0$. (24) now follows from (25), (27) and (28) in the case $\ell = 0$.

Now suppose that $\ell > 0$ and the theorem has been proved for all functions of $W^{\ell-1,p}(R^n)$. Let $f \in W^{\ell,p}(R^n)$. By the Poincare inequality,

$$\begin{aligned} &\rho^{(k-\ell)p-n} \int_{|y-x|<\rho} |f(y)-f_{p,x}|^p \, dy \\ &\leq C\rho^{(k-(\ell-1))p-n} \sum_{i=1}^{n} \int_{|y-x|<\rho} \left|\frac{\partial f}{\partial y_i}(y)\right|^p dy, \end{aligned} \tag{29}$$

for all $x \in R^n$, where C depends only on n. By the induction assumption, there exists a set F', with

$$R_{k,p}(F') = 0 \tag{30}$$

and

$$\rho^{(k-(\ell-1))p-n} \int_{|y-x|<\rho} \left|\frac{\partial f}{\partial y_i}(y) - \left(\frac{\partial f}{\partial y_i}\right)_{p,x}\right|^p dy \to 0 \tag{31}$$

as $\rho \to 0+$, for all $x \in R^n \sim F'$. But

$$\begin{aligned} \left[\int_{|y-x|<\rho} \left|\frac{\partial f}{\partial y_i}(y)\right|^p dy\right]^{\frac{1}{p}} &\leq \left[\int_{|y-x|<\rho} \left|\frac{\partial f}{\partial y_i}(y) - \left(\frac{\partial f}{\partial y_i}\right)_{p,x}\right|^p dy\right]^{\frac{1}{p}} \\ &+ \left|\left(\frac{\partial f}{\partial y_i}\right)_{\rho,x}\right| \left[\int_{|y-x|<\rho} dy\right]^{\frac{1}{p}}, \end{aligned} \tag{32}$$

and by 3.6

$$\rho^{k-(\ell-1)} \left(\frac{\partial f}{\partial y_i}\right)_{\rho,x} \to 0 \tag{33}$$

as $\rho \to 0+$, for all $x \in R^n$, except for a set F'' with

$$R_{k,p}(F'') = 0 .$$

(24) now follows from (29), (31), (32) and (33). This completes the proof.

3.8 THEOREM

Let ℓ *be a positive integer such that* $\ell p < n$, *let* Ω *be an open set of* R^n *and let* $f \in W^{\ell,p}(\Omega)$ *be approximately continuous except for a set* E' *with* $R_{\ell,p}(E') = 0$. *Then*

$$\rho^{-n} \int_{|y-x|<\rho} |f(y)-f(x)|^p \, dy \to 0 \tag{34}$$

as $\rho \to 0+$, *for all* $x \in \Omega$, *except for a set* E *with* $R_{\ell,p}(E) = 0$.

This generalizes a result of Federer and Ziemer in [4] as described in the introduction.

PROOF: (i) When $\Omega = R^n$, (34) follows from 3.7 and 3.4.

(ii) When Ω is arbitrary, the theorem can be derived from (i) as in the proof of 3.4.

3.9 COROLLARY

Let ℓ *be a positive integer such that* $\ell p < n$, *let* Ω *be an open set of* R^n *and let* $f \in W^{\ell,p}(\Omega)$ *be approximately continuous except for a set* E', *with* $R_{\ell,p}(E') = 0$. *Then*

$$\lim_{\rho\to 0+} \rho^{-n} \int_{|y-x|<\rho} |f(y)|^p \, dy \quad \textit{exists and} \quad = |f(x)|^p \tag{35}$$

for all $x \in \Omega$, *except for a set* E *with* $R_{\ell,p}(E) = 0$.

Our next theorem gives an expansion of Taylor type for Sobolev functions. When ℓ, m are integers such that $0 \le m \le \ell$, $(\ell-m)p < n$ and $f \in W^{\ell,p}(R^n)$, it follows from 3.4 that there exists a subset E of R^n such that

$$R_{\ell-m,p}(E) = 0 \tag{36}$$

and

$$\lim_{\rho\to 0+} \frac{1}{m(B_\rho)} \int_{|y-x|<\rho} D^\alpha f(y)dy \tag{37}$$

exists for all $x \in R^n \sim E$ and every multi-index α with $0 \le |\alpha| \le m$. When the limit in (37) exists, we denote it by $D^\alpha f(x)$.

When $x \in R^n \sim E$, we denote by $T^{(m)}(f,x;\cdot)$, the Taylor polynomial, defined by

$$T^{(m)}(f,x;y) = \sum_{0\le|\alpha|\le m} \frac{1}{(\alpha_1!)(\alpha_2!)\cdots\cdots(\alpha_n!)}$$

$$\cdot\ D^\alpha f(x)(y_1-x_1)^{\alpha_1}(y_2-x_2)^{\alpha_2}\cdots\cdots(y_n-x_n)^{\alpha_n} . \tag{38}$$

We observe that when f is a C^m function on R^n, Taylor's theorem can be expressed in the form

$$f(y) = T^{(m-1)}(f,x;y) + m \sum_{|\alpha|=m} \frac{1}{(\alpha_1!)(\alpha_2!)\cdots\cdots(\alpha_n!)}$$

$$\cdot \left[\int_0^1 (1-t)^{m-1} D^\alpha f((1-t)x+(y)dt\right](y_1-x_1)^{\alpha_1}(y_2-x_2)^{\alpha_2}\cdots\cdots$$

$$\cdots\cdots(y_n-x_n)^{\alpha_n} \tag{39}$$

The Taylor expansion, given here, differs slightly from the one given by Meyers in [9]. The expansion in the present paper is valid for $1 \le p < \infty$ and has the remainder in a different form to that of Meyers.

3.10 TAYLOR's THEOREM

Let $1 \le m \le \ell$ *and suppose* $(\ell-m)p < n$. *Let* $f \in W^{\ell,p}(R^n)$ *and* E *be the set described in* (36) *and* (37). *Then*

$$\left[\int_{|y-x|<\rho} |f(y)-T^{(m)}(f,x;y)|^p\, dy\right]^{\frac{1}{p}} \le \rho^m \sum_{|\alpha|=m} \frac{m}{(\alpha_1!)(\alpha_2!)\cdots\cdots(\alpha_n!)}$$

$$\cdot \int (1-t)^{m-1} t^{-\frac{n}{p}} \left[\int_{|y-x|<t\rho} |D^\alpha f(y)-D^\alpha f(x)|^p\, dy\right]^{\frac{1}{p}} dt \tag{40}$$

and

$$\left[\int_{|y-x|<\rho} |f(y)-T^{(m-1)}(f,x;y)|^p\, dy\right]^{\frac{1}{p}} \le \rho^m \sum_{|\alpha|=m} \frac{m}{(\alpha_1!)(\alpha_2!)\cdots(\alpha_n!)}$$

$$\cdot \int_0^1 (1-t)^{m-1} t^{-\frac{n}{p}} \left[\int_{|y-x|<t\rho} |D^\alpha f(y)|^p\, dy\right]^{\frac{1}{p}} dt , \tag{41}$$

for all $x \in R^n$ *except for a set* E', *with* $E' \supset E$ *and* $R_{\ell-m,p}(E') = 0$.

PROOF: (i) Suppose first of all that f is a C^m function on R^n. Let $q = \frac{p}{p-1}$ when $p > 1$ and $=\infty$ when $p = 1$. Let $x \in R^n$, $\rho > 0$ and put $B = B_\rho(x)$. Let ϕ be a function of $L^q(B)$ with $|\phi| \le 1$. By (38) and (39),

$$\int_B [f(y)-T^{(m)}(f,x;y)]\phi(y)dy = \sum_{|\alpha|=m} \frac{m}{(\alpha_1!)(\alpha_2!)\ldots(\alpha_n!)}$$

$$\cdot \int_0^1 (1-t)^{m-1}\Big[\int_B \{D^\alpha f((1-t)x+ty)-D^\alpha f(x)\}(y_1-x_1)^{\alpha_1}\ldots(y_n-x_n)^{\alpha_n}\phi(y)dy\Big]dt ,$$

hence

$$\left|\int_B [f(y)-T^{(m)}(f,x;y)]\phi(y)dy\right| \le \rho^m \sum_{|\alpha|=m} \frac{m}{(\alpha_1!)\ldots(\alpha_n!)}$$

$$\cdot \int_0^1 (1-t)^{m-1}\left[\int_B |D^\alpha f(1-t)x+ty)-D^\alpha f(x)|^p\, dy\right]^{\frac{1}{p}} dt .$$

By making the substitution $z = x + t(y-x)$ in the right-hand side and then taking the supremum over all ϕ, one obtains (40).

The inequality (41) can be derived similarly.

(ii) Now let f be an arbitrary function of $W^{\ell,p}(R^n)$. By 3.9, there exists a set $E' \supset E$, with $R_{\ell-m,p}(E') = 0$ and such that

$$\lim_{\delta\to 0+} \delta^{-n} \int_{|y-x|<\delta} |D^\alpha f(y)|^p\, dy \quad \text{exists and} \quad = |D^\alpha f(x)|^p , \tag{42}$$

when $|\alpha| = m$ and $x \in R^n \sim E'$.

Consider an $x \in R^n \sim E'$. There exists a constant M (depending on x), such that

$$\delta^{-n} \int_{|y-x|<\delta} |D^\alpha f(y)|^p\, dy \le M \tag{43}$$

for all $|\alpha| = m$ and all $\delta > 0$. There now exists a sequence $\{\eta_s\}$ (depending on f and x) of non-negative C^∞ functions on R^n, such that

$$\int_{R^n} \eta_s(x)dx = 1 , \tag{44}$$

spt $\eta_s \subset B_{\frac{1}{s}}(0)$ and

$$\eta_s \le \Lambda s^n \tag{45}$$

for all s (where Λ depends only on n), while

$$(\eta_s * D^\alpha f)(x) \to D^\alpha f(x) \tag{46}$$

as $s \to \infty$, for $0 \le |\alpha| \le m$. Put $f_s = \eta_s * f$. Each $f_s \in C^\infty(R^n) \cap W^{\ell,p}(R^n)$, so that by (i), the inequalities (40) and (41) hold when f is replaced by f_s. The required inequalities will follow from standard mollifier theory when we show that for each α with $|\alpha| = m$,

$$t^{-n} \int_{|y-x|<t\rho} |D^\alpha f_s(y)|^p \, dy \tag{47}$$

has a bound which is independent of t and s.

Now, for any measurable subset B of R^n, we have (when $|\alpha| = m$)

$$\int_B |D^\alpha f_s(y)|^p \, dy = \int_B \left| \int_{R^n} \eta_s(y-z) D^\alpha f(z) dz \right|^p dy ,$$

hence by (45)

$$\int_B |D^\alpha f_s(y)|^p \, dy \le \Lambda^p s^{np} \int_B \left[\int_{|y-z|<\frac{1}{s}} |D^\alpha f(z)| dz \right]^p dy . \tag{48}$$

Thus, when $p > 1$

$$\int_B |D^\alpha f_s(y)|^p \, dy \le \Lambda^p s^{np} \int_B \left[\int_{|y-z|<\frac{1}{s}} |D^\alpha f(z)|^p \, dz \right] \left[\int_{|y-z|<\frac{1}{s}} dz \right]^{p-1} dy ,$$

so that

$$\int_B |D^\alpha f_s(y)|^p \, dy \le A s^n \int_B \int_{|z-y|<\frac{1}{s}} |D^\alpha f(z)|^p \, dz \, dy , \tag{49}$$

where A depends only on n and p. When $p = 1$, (49) follows from (48).

When $t\rho \le 3s^{-1}$, we let B be the disc with centre x and radius $t\rho$. By (49)

$$\int_{|y-x|<t\rho} |D^\alpha f_s(y)|^p \, dy \le A s^n \int_{|y-x|<t\rho} \int_{|z-x|<4s^{-1}} |D^\alpha f(z)|^p \, dz \, dy .$$

It now follows from (43) that (47) holds in the case where $t\rho \le 3s^{-1}$.

When $t\rho > 3s^{-1}$, we have

$$\int_{|y-x|<t\rho} |D^\alpha f_s(y)|^p \, dy = \int_{|y-x|<3s^{-1}} |D^\alpha f_s(y)|^p \, dy + \int_{3s^{-1} \le |y-x| < t\rho} |D^\alpha f_s(y)|^p \, dy$$

and a double application of (49) yields

$$\int_{|y-x|<t\rho} |D^\alpha f_s(y)|^p\,dy \le As^n \int_{|y-x|<3s^{-1}} \int_{|z-x|<4s^{-1}} |D^\alpha f(z)|^p\,dz\,dy$$

$$+ As^n \int_{3s^{-1}\le|y-x|<t\rho} \int_{|z-y|<s^{-1}} |D^\alpha f(z)|^p\,dz\,dy$$

and by (43)

$$\le A't^n\rho^n + As^n \int_{3s^{-1}\le|y-x|<t\rho} \int_{|w|<s^{-1}} |D^\alpha f(w+y)|^p\,dw\,dy$$

$$\le A't^n\rho^n + As^n \int_{|w|<s^{-1}} \left[\int_{2s^{-1}\le|u-x|<t\rho+s^{-1}} |D^\alpha f(y)|^p\,du\right] dw$$

so that

$$\int_{|y-x|<t\rho} |D^\alpha f_s|^p\,dy \le A''t^n\rho^n\ .$$

Thus (47) holds in this case also.

This completes the proof.

3.11 THEOREM

Let $0 \le m \le \ell$ *and suppose* $(\ell-m)p < n$. *Let* $f \in W^{\ell,p}(R^n)$. *Then*

$$\rho^{-m}\left[\rho^{-n} \int_{|y-x|<\rho} |f(y)-T^{(m)}(f,x;y)|^p\,dy\right]^{\frac{1}{p}} \to 0$$

as $\rho \to 0+$, *for all* $x \in R^n$, *except for a set* F *with*

$$R_{\ell-m,p}(F) = 0\ .$$

PROOF: When $m = 0$, the theorem reduces to 3.8. Suppose $m > 0$. By 3.8,

$$\rho^{-n} \int_{|y-x|<\rho} |D^\alpha f(y)-D^\alpha f(x)|^p\,dy \to 0 \tag{50}$$

as $\rho \to 0+$, for all $|\alpha| = m$ and all $x \in R^n$, except for a set F with

$$R_{\ell-m,p}(F) = 0\ .$$

Consider an $x \in R^n \sim F$ and an α with $|\alpha| = m$. Define

$$\eta(\lambda) = \left[\lambda^{-n} \int_{|y-x|<\lambda} |D^\alpha f(y) - D^\alpha f(x)|^p \, dy\right]^{\frac{1}{p}} \tag{51}$$

for $\lambda > 0$. By (50), $\eta(\lambda) \to 0$ as $\lambda \to 0+$, hence

$$\int_0^1 (1-t)^{m-1} \eta(t\rho)dt \to 0 \tag{52}$$

as $\rho \to 0+$. The required result now follows from (51), (52) and 3.10.

3.12 LEMMA

Let ℓ *be a positive integer and* k *a real number such that* $k \geq 0$ *and* $(k+\ell-1)p < n$. *There exists a constant* C *depending only on* k, p, n *and* ℓ *and such that*

$$\delta^{k-n} \left| \int_{|y-x|<\delta} f(y)dy \right| \leq C \sum_{|\alpha|=\ell} \int_{R^n} |y-x|^{k-n+\ell} |D^\alpha f(y)|dy + \delta^k |f|_p \tag{53}$$

for all $f \in W^{\ell,p}(R^n)$ *every real number* k *all* $\delta \in (0,1]$ *and all* $x \in R^n$.

PROOF: Let f, k, δ and x be as described. By 3.1 (with $\rho = 1$)

$$\delta^{k-n} \left| \int_{|y-x|<\delta} f(y)dy \right| \leq \delta^k \int_{|y-x|<1} |f(y)|dy + \frac{1}{n} \delta^{k-n} \int_{|y-x|<\delta} |y-x||Df(y)|dy$$

$$+ \frac{1}{n} \delta^k \int_{|y-x|<1} |y-x||Df(y)|dy + \frac{1}{n} \delta^k \int_{\delta<|y-x|<1} |y-x|^{1-n} |Df(y)|dy . \tag{54}$$

But

$$\frac{1}{n} \delta^{k-n} \int_{|y-x|<\delta} |y-x||Df(y)|dy \leq \frac{1}{n} \int_{|y-x|<\delta} |y-x|^{k-n+1} |Df(y)|dy \tag{55}$$

and

$$\frac{1}{n} \delta^k \int_{|y-x|<1} |y-x||Df(y)|dy \leq \frac{1}{n} \int_{|y-x|<1} |y-x||Df(y)|dy ,$$

so that

$$\frac{1}{n} \delta^k \int_{|y-x|<1} |y-x||Df(y)|dy \leq \frac{1}{n} \int_{|y-x|<1} |y-x|^{k-n+1} |Df(y)|dy . \tag{56}$$

Also

$$\frac{1}{n} \delta^k \int_{\delta<|y-x|<1} |y-x|^{+n} |Df(y)|dy = \frac{1}{n} \int_{\delta<|y-x|<1} |y-x|^{k-n+1} |Df(y)| . \tag{57}$$

By (54), (55), (56) and (57),

$$\delta^{k-n}\left|\int_{|y-x|<\delta} f(y)dy\right| \leq \delta^k |f|_p + \frac{3}{n}\int_{R^n} |y-x|^{k-n+1}|Df(y)|dy . \tag{58}$$

This is the inequality (53) in the case $\ell = 1$.

Suppose $\ell > 1$. By 3.3,

$$\int_{R^n} |y-x|^{k-n+1}\left|\frac{\partial f}{\partial y_i}(y)\right|dy \leq C \sum_{|\alpha|=\ell} \int_{R^n} |y-x|^{k-n+\ell}|D^\alpha f(y)|dy . \tag{59}$$

The inequality (53) now follows from (58) and (59).

3.13 LEMMA

Let ℓ *be a positive integer such that* $\ell p < n$ *and let* k *be a real number such that* $\ell \leq k < \frac{n}{p}$. *Let* $f \in W^{\ell,p}(R^n)$ *and* $\varepsilon > 0$. *There exists an open set* U *and a constant* M , *such that*

$$R_{k,p}(U) < \varepsilon$$

and

$$\delta^{k-\ell-n}\left|\int_{|y-x|<\delta} f(y)dy\right| \leq M$$

for all $x \in R^n \sim U$ *and* $\delta \in (0,1]$.

PROOF: Define

$$h(z) = \sum_{|\alpha|=\ell} |D^\alpha f(z)|$$

for $z \in R^n$. Let Γ be a positive real number and put

$$h_\Gamma = \frac{1}{\Gamma} h .$$

Then the set

$$F_\Gamma = \{x;\ x \in R^n \text{ and } I_k h_\Gamma(x) \geq 1\}$$

has

$$R_{k,p}(F_\Gamma) \leq |h_\Gamma|_p^p = \frac{1}{\Gamma^p}|h|_p^p ,$$

so that

$$R_{k,p}(F_\Gamma) \le \frac{1}{\Gamma^p}\left(\sum_{|\alpha|=\ell} |D^\alpha f|_p\right)^p . \tag{60}$$

When $x \notin F_\Gamma$, we have

$$\frac{1}{\Gamma_\gamma(k)} \sum_{|\alpha|=\ell} \int_{R^n} |y-x|^{k-n} |D^\alpha f(y)|dy < 1$$

and hence by 3.12, with $k - \ell$ substituted for k ,

$$\delta^{k-\ell-n}\left|\int_{|y-x|<\delta} f(y)dy\right| \le C\Gamma_\gamma(k) + \delta^{k-\ell}|f|_p \tag{61}$$

The existence of the open set U now follows from 2(B), (60) and (61).

3.14 COROLLARY

Let ℓ *be a non-negative integer, let* $f \in W^{\ell,p}(R^n)$ *and suppose that* f *is approximately continuous on* R^n *except for a set* E *with* $R_{\ell,p}(E) = 0$. *Let* $\varepsilon > 0$. *There exists an open set* U *of* R^n *and a constant* M *such that* $U \supset E$,

$$R_{\ell,p}(U) < \varepsilon$$

and

$$|f(x)| \le M$$

for all $x \in R^n \sim U$.

3.15 LEMMA

Let ℓ *be a non-negative integer such that* $\ell p < n$ *and* k *a real number such that* $\ell \le k < \frac{n}{p}$. *Let* $f \in W^{\ell,p}(R^n)$ *and let* $\varepsilon > 0$. *There exists an open set* U *and a constant* M *such that*

$$R_{k,p}(U) < \varepsilon \tag{62}$$

and

$$\delta^{(k-\ell)p-n} \int_{|y-x|<\delta} |f(y)|^p \, dy \le M \tag{63}$$

for all $x \in R^n \sim U$ *and* $\delta > 0$.

PROOF: By induction on ℓ .

(i) Suppose $\ell = 0$. For each Borel subset B of R^n , define

$$\mu(B) = \int_B |f(y)|^p \, dy \; .$$

Then μ is a finite measure.

By taking a sufficiently large σ in Lemma 3.2 of [1] and applying 2(F) and 2(B) one can obtain the open set U in this case.

(ii) Now suppose $\ell > 0$ and there exists an open set U' and a constant M' such that

$$R_{k,p}(U') < \frac{1}{2}\varepsilon \tag{64}$$

and

$$\delta^{(k-\ell+1)p-n} \int_{|y-x|<\delta} \left|\frac{\partial f}{\partial y_i}(y)\right|^p dy \le M' \tag{65}$$

for $i = 1,\ldots,n$, $x \in R^n \sim U'$ and $\delta > 0$. By the Poincare inequality

$$\int_{|y-x|<\delta} |f(y)-f_{\delta,x}|^p \, dy \le C\delta^p \int_{|y-x|<\delta} |Df(y)|^p \, dy$$

for all $x \in R^n$, so that by (65)

$$\delta^{(k-\ell)p-n} \int_{|y-x|<\delta} |f(y)-f_{\delta,x}|^p \, dy \le CnM' \; , \tag{66}$$

for $x \in R^n \sim u'$ and $\delta > 0$. Now

$$\delta^{(k-\ell)p-n} \int_{|y-x|<\delta} |f(y)|^p \, dy \le 2^p \delta^{(k-\ell)p-n} \int_{|y-x|<\delta} |f(y)-f_{\delta,x}|^p \, dy + 2^p \delta^{(k-\ell)p-n} \int_{|y-x|<\delta} |f_{\delta,x}|^p \, dy \tag{67}$$

for $x \in R^n \sim U'$ and $\delta > 0$. But by 3.13 there exists an open set U'' and a constant M' such that

$$R_{k,p}(U'') < \frac{1}{2}\varepsilon \tag{68}$$

and

$$\delta^{k-\ell} |f_{\delta,x}| \le M'' \; , \tag{69}$$

for $x \in R^n \sim U''$ and $\delta \in (0,1]$. Then

$$2^{p}\delta^{(k-\ell)p-n} \int_{|y-x|<\delta} |f_{\delta,x}|^{p}\, dy \le M''' \tag{70}$$

for $x \in R^n \sim U'''$ and $\delta \in (0,1]$, where M''' is a constant. Put $U = U' \cup U''$. Then U satisfies (62). It follows from (66), (67) and (70) that (63) is satisfied when $x \in R^n \sim U$ and $\delta \in (0,1]$. Clearly there exists a constant M such that (63) is satisfied when $\delta \in (1,\infty)$.

This completes the induction.

3.16 THEOREM

Let ℓ, m *be positive integers such that* $m \le \ell$ *and* $(\ell-m)p < n$. *Let* $f \in W^{\ell,p}(R^n)$ *and suppose that* f *and all of its partial derivatives of order* $\le m$ *are approximately continuous everywhere except for a set* E *with* $R_{\ell-m,p}(E) = 0$. *Let* $\varepsilon > 0$. *Then there exists an open set* U *of* R^n *and a* C^m *function* g *on* R^n, *such that* $U \supset E$,

$$R_{\ell-m,p}(U) < \varepsilon$$

and

$$D^{\alpha}g(x) = D^{\alpha}f(x)$$

for all $x \in R^n \sim U$ *and* $0 \le |\alpha| \le m$.

PROOF: By substituting $\ell - m$ for k and ℓ in 3.15 and using 2(B), one can obtain an open set $U_1 \supset E$ and a constant M_1, such that

$$R_{\ell-m,p}(U_1) < \frac{1}{3}\varepsilon$$

and

$$\delta^{-n} \int_{|y-x|<\delta} |D^{\alpha}f(y)|^{p}\, dy \le M_1$$

for $x \in R^n \sim U_1$, $|\alpha| = m$ and $\delta > 0$. It now follows from (41) that

$$\left[\delta^{-n} \int_{|y-x|<\delta} |f(y)-T^{(m-1)}(f,x;y)|^{p}\, dy\right]^{\frac{1}{p}} \le M_1' \delta^{m}$$

for $x \in R^n \sim U_1$ and $\delta > 0$, where M_1' is a constant. By 3.14 there exists an open set $U_2 \supset E$ and a constant M_2 such that

$$R_{\ell-m,p}(U_2) < \frac{1}{3}\varepsilon$$

and the coefficients of $T^{(m-1)}(f,x;\cdot)$ are bounded by M_2 for all $x \in R^n \sim U_2$. By 2(B) and 3.11, there exists an open set $U_3 \supset E$ and such that

$$R_{\ell-m,p}(U_3) < \frac{1}{3}\varepsilon$$

and

$$\delta^{-m}\left[\delta^{-n}\int_{|y-x|<\delta} |f(y)-T^{(m)}(f,x;y)|^p \, dy\right]^{\frac{1}{p}} \to 0$$

as $\delta \to 0+$, for all $x \in R^n \sim U_3$. By putting

$$U = U_1 \cup U_2 \cup U_3$$

and applying Theorem 9 of [3], one can obtain the required open set U and the function g.

4 THE MAIN APPROXIMATION

In 4 we prove the main approximation theorem for Sobolev functions. In addition to some preliminary lemmas, we will need the Poincare' inequality in the following form.

4.1 THEOREM

Let $\sigma \in (0,1)$ *and let* ℓ *be a positive integer. There exists a constant* C *depending only on* σ, ℓ, p *and* n *and such that:*

for every non-empty bounded convex open subset Ω *of* R^n *with diameter* ρ *and every* $f \in W^{\ell,p}(\Omega)$ *for which the set*

$$\{x;\ x \in \Omega \text{ and } f(x) = 0\}$$

has measure $\geq \sigma m(\Omega)$ *the inequality*

$$\int_\Omega |f(x)|^p \, dx \leq C\rho^{\ell p} \sum_{|\alpha|=\ell} \int_\Omega |D^\alpha f(x)|^p \, dx$$

holds.

This form of the Poincare' inequality is easily derived from the inequality (7.45) on page 157 of [5].

4.2 LEMMA

Let ℓ *be a positive integer and let* f *be a function of* $W^{\ell,p}(R^n)$ *which vanishes outside a bounded open set* U.

Let δ , $\sigma \in (0,1)$ *and let* E *be the set of all those points* x *of* ∂U *such that*

$$\inf_{0<t\leq\delta} t^{-n} m(K_t(x)\cap(\sim U)) \geq \sigma , \tag{71}$$

where $K_t(x)$ *denotes the closed cube with centre* x *and edge-length* t .

Let m *be a positive integer such that* $m \leq \ell$ *and let* $\varepsilon > 0$.

Then there exists a function $g \in W^{m,p}(R^n)$ *and an open set* V *such that*

(i) $|f-g|_{m,p} < \varepsilon$;

(ii) $E \subset V$ *and* $g(x) = 0$ *when* $x \in (\sim U) \cup V$.

PROOF: For $\lambda \in (0,1]$, let K_λ denote the set of all closed cubes of the form

$$[(i_1-1)\lambda, i_1\lambda]\times[(i_2-1)\lambda, i_2\lambda]\times\ldots\ldots\times[(i_n-1)\lambda, i_n\lambda] ,$$

where i_1 , $i_2,\ldots.,i_n$ are arbitrary integers. Consider a $\lambda < \frac{1}{3}\delta$ and let

$$K^{(1)}, K^{(2)},\ldots.,K^{(r)}$$

be those cubes of K_λ that intersect E . Let $a^{(i)}$ be the centre of $K^{(i)}$ and let

$$P^{(i)} = K_{4\lambda}(a^{(i)}) .$$

Let ζ be a C^∞ function on R^n , such that $0 \leq \zeta \leq 1$, $\zeta(x) = 0$, when $x \in K_1(0)$ and $\zeta(x) = 1$ when $x \notin K_{\frac{3}{2}}(0)$. Define

$$g_\lambda(x) = f(x) \prod_{i=1}^{r} \zeta(\tfrac{1}{2}\lambda^{-1}(x-a^{(i)})) \tag{72}$$

for $x \in R^n$. Clearly $g_\lambda(x) = 0$ when $d(x,E) \leq \frac{1}{2}\lambda$, so that, for any λ , we can put $g = g_\lambda$ and find an open set V satisfying (ii).

We keep i fixed for the moment and estimate

$$|f-g_\lambda|_{\ell,p,P^{(i)}} . \tag{73}$$

We observe that there exists a constant τ , depending only on n and such that at most τ of the cubes $P^{(j)}$ intersect $P^{(i)}$ (including $P^{(i)}$) . Denote these by

$$P^{(j_1)}, P^{(j_2)}, \ldots\ldots, P^{(j_s)}$$

where $s \le \tau$. Then, for $x \in P^{(i)}$

$$q_\lambda(x) = f(x)w(x) \ , \tag{74}$$

where

$$w(x) = \prod_{k=1}^{s} \zeta(\tfrac{1}{2}\lambda^{-1}(x-a^{(j_k)})) \ . \tag{75}$$

Now, for $x \in P^{(i)}$ and any multi-index α with $0 \le |\alpha| \le \ell$, we have

$$|D^\alpha w(x)| \le A_1 \lambda^{-|\alpha|} \ ,$$

where A_1 depends only on ℓ and n . Hence for almost all $x \in P^{(i)}$ and any multi-index γ with $0 \le |\gamma| \le \ell$, we have

$$|D^\gamma g_\lambda(x)| \le A_2 \sum_{r=0}^{|\gamma|} \lambda^{r-|\gamma|} \sum_{|\beta|=r} |D^\beta f(x)| \ , \tag{76}$$

where A_2 depends only on ℓ and n .

Let y be a point where $K^{(i)}$ intersects E . Clearly, there is a subcube $Q^{(i)}$ of $P^{(i)}$ with centre y and edges of length 3λ . By (71), f and hence its derivatives are zero on a subset Z of $Q^{(i)}$ with

$$m(Z) \ge \sigma(3\lambda)^n \ . \tag{77}$$

By applying the Poincare' inequality to the interior of the convex set $P^{(i)}$ we obtain, when $|\beta| < \ell$,

$$\int_{P^{(i)}} |D^\beta f(x)|^p \, dx \le A_3 \lambda^{p(\ell-|\beta|)} \sum_{|\xi|=\ell} \int_{P^{(i)}} |D^\xi f(x)|^p \, dx \ , \tag{78}$$

where A_3 depends only on ℓ , n , σ and p . But, with a suitable constant A_3 , (78) will still hold when $|\beta| = \ell$. By (76) and (78) (since $\lambda \le 1$)

$$\int_{P^{(i)}} |D^\gamma g_\lambda(x)|^p \, dx \le A_4 \sum_{|\xi|=\ell} \int_{P^{(i)}} |D^\xi f(x)|^p \, dx \tag{79}$$

for $0 \le |\gamma| \le \ell$, where A_4 depends only on ℓ , n , p and σ . Put

$$X_\lambda = \bigcup_{i=1}^{r} P^{(i)} \ .$$

Then

$$\int_{X_\lambda} |D^\gamma g_\lambda(x)|^p \, dx \le A_4 \sum_{|\xi|=\ell} \sum_{i=1}^{r} \int_{\rho(i)} |D^\xi f(x)|^p \, dx$$

for $0 \le |\gamma| \le \ell$. But each point of X_λ belongs to at most τ of the cubes $P^{(i)}$, hence

$$\int_{X_\lambda} |D^\gamma g_\lambda(x)|^p \, dx \le \tau A_4 \sum_{|\xi|=\ell} \int_{X_\lambda} |D^\xi f(x)|^p \, dx , \tag{80}$$

for $0 \le |\gamma| \le \ell$. Now

$$|f-g_\lambda|^p_{\ell,p} \le 2^p \sum_{0\le|\gamma|\le\ell} \left[\int_{X_\lambda} |D^\gamma g_\lambda(x)|^p \, dx + \int_{X_\lambda} |D^\gamma f(x)|^p \, dx \right] ,$$

so that by (80)

$$|f-g_\lambda|^p_{\ell,p} \le A_5 \sum_{0\le|\gamma|\le\ell} \int_{X_\lambda \cap U} |D^\gamma f(x)|^p \, dx , \tag{81}$$

where A_5 depends only on ℓ, n, p and σ. But $X_\lambda \cap U \subset \{x;\ x \in U \text{ and } d(x,\partial U) < 3n^{\frac{1}{2}}\lambda\}$, hence $m(X_\lambda \cap U) \to 0$ as $\lambda \to 0+$. Therefore by (81)

$$|f-g_\lambda|_{\ell,p} \to 0$$

as $\lambda \to 0+$.

The required function g is now obtained by putting $g = g_\lambda$, with sufficiently small.

4.3 LEMMA

Let λ *be a positive real number which is* $<n$. *There exists a constant* C, *depending only on* λ *and* n *and such that*

$$\delta^{-n} \int_{|x-z|<\delta} |x-y|^{\lambda-n} \, dx \le C|y-z|^{\lambda-n} , \tag{82}$$

for all $y, z \in R^n$ *and all* $\delta > 0$.

PROOF: We show first of all that there exists a constant C, such that (82) holds when $y = 0$ and z is arbitrary.

When $|z| \ge 3\delta$, we have

$$|z| \le |x| + |z-x| < |x| + \delta \le |x| + \tfrac{1}{3}|z| ,$$

so that $|z| \le \frac{3}{2}|x|$ and $|x|^{\lambda-n} \le A|z|^{\lambda-n}$, so that (82) holds.

When $|z| < 3\delta$,

$$\delta^{-n} \int_{|x-z|<\delta} |x|^{\lambda-n}\, dx \le \delta^{-n} \int_{|x|<4\delta} |x|^{\lambda-n}\, dx .$$

By evaluating the integral on the right, one can again see that (82) holds with $y = 0$.

Since we have shown that

$$\delta^{-n} \int_{|x-z|<\delta} |x|^{\lambda-n}\, dx \le C|z|^{\lambda-n} ,$$

for all $z \in R^n$, we can replace z by $z - y$ and then make the substitution $x = w - y$, dsc = div in the integral, to obtain (82).

4.4 LEMMA

Let k *be a non-negative real number such that* $kp < n$. *There exists a constant* C *depending only on* n, k *and* p *and such that:*

for all $\sigma \in (0,1)$ *every bounded non-empty open subset* U *of* R^n *and every subset* F *of* ∂U *with the property that for each* $x \in F$, *there is a* $t \in (0,1)$ *for which*

$$\frac{m(U \cap B_t(x))}{m(B_t(x))} \ge \sigma , \tag{83}$$

the inequality

$$R_{k,p}(U \cup F) \le C\sigma^{-p} R_{k,p}(U) \tag{84}$$

holds.

PROOF: Let σ, U and F be as described above. The cases $k = 0$ and $k > 0$ are treated separately.

(i) We consider first the case where $k > 0$. Let ψ be a non-negative function of $L_p(R^n)$ such that

$$\frac{1}{\gamma(k)} \int_{R^n} |x-y|^{k-n}\, \psi(y) dy \ge 1 \tag{85}$$

for all $x \in U$. Let C_1 be the constant of 4.3. It can be assumed that $C_1 \ge 1$. Consider a point $b \in F$ and let t be such that (83) holds for $x = b$. By 4.3

$$C_1 |y-b|^{k-n} \geq t^{-n} \int_{|x-b|<t} |x-y|^{k-n} \, dx$$

so that

$$C_1 \int_{R^n} |y-b|^{k-n} \psi(y)dy \geq t^{-n} \int_{|x-b|<t} \left[\int_{R^n} |x-y|^{k-n} \psi(y)dy \right] dx$$

and by (85),

$$\geq \gamma(k) t^{-n} \, m(U \cap B_t(b)) \; .$$

Hence by (83),

$$\frac{C_2}{\gamma(k)} \int_{R^n} |y-b|^{k-n} \psi(y)dy \geq \sigma \; .$$

Put $\eta = \frac{C_2}{\sigma} \psi$. Then

$$\frac{1}{\gamma(k)} \int_{R^n} |y-x|^{k-n} \, \eta(y)dy \geq 1$$

for all $x \in F$, so that

$$R_{k,p}(F) \leq |\eta|_p^p = \left(\frac{C_2}{\sigma}\right)^p |\psi|_p^p \; .$$

Thus

$$R_{k,p}(F) \leq \left(\frac{C_2}{\sigma}\right)^p R_{k,p}(U) \; .$$

The required inequality now follows.

(ii) Now let $k = 0$, so that $R_{k,p} = m^*$. Let $\mathcal{B}$ be the collection of all those closed balls B with centre $\in F$, radius $\in (0,1)$ and such that

$$\frac{m(U \cap B)}{m(B)} \geq \sigma \; . \tag{86}$$

Clearly, conditions (i) and (ii) of 3.5 are satisfied. Since the union of the sets of $\mathcal{B}$ is contained in a bounded set, it follows that condtion (iii) of 3.5 is also satisfied. Hence, there exists a sequence $\{B_r\}$ in $\mathcal{B}$, such that $B_r \cap B_s = \emptyset$ when $r \neq s$ and

$$F \subset \bigcup_r B_r^* \; .$$

Thus

$$m^*(F) \le \sum_r m(B_r^*) = 5^n \sum_r m(B_r) .$$

and by (86)

$$\le 5^n \sigma^{-1} \sum_r m(U \cap B_r) \le 5^n \sigma^{-1} m(U) .$$

Since $\sigma < 1$, the required inequality follows from this.

4.5 LEMMA

Let k *be a non-negative real number such that* $kp < n$ *and let* ℓ *be a positive integer. There exists a constant* C , *depending only on* n , p , k *and* ℓ *and such that:*

for every bounded non-empty open subset U *of* R^n *every* $f \in W^{\ell,p}(R^n)$ *which vanishes outside* U *and every* $\varepsilon > 0$ *there exists a* C^∞ *function* h *on* R^n *with the properties*

(i) $|f-h|_{\ell,p} < \varepsilon,$

(ii) $R_{k,p}(\text{spt } h) \le CR_{k,p}(U)$ and

(iii) spt h *is contained in the set*

$$V = \{x;\ x \in R^n \text{ and } d(x,U) < \varepsilon\} .$$

PROOF: Let U , f and ε be as described above. Since $U \neq \emptyset$, it follows that $R_{k,p}(U) > 0$. Denote by E , the set of all those points x of ∂U such that

$$\inf_{0<t\le n^{-\frac{1}{2}}} t^{-n} m(K_t(x) \cap (\sim U)) \ge \tfrac{1}{2} , \tag{87}$$

where $K_t(x)$ is the closed cube with centre x and edge length t . Then E is closed. By 4.2 there exists a function $g \in W^{\ell,p}(R^n)$ and an open set V_0 such that

$$|f-g|_{\ell,p} < \tfrac{1}{2}\varepsilon , \tag{88}$$

$E \subset V_0$ and $g(x) = 0$ when $x \in (\sim U) \cup V_0$. Put

$$F = \partial U \sim E .$$

Then there is a $\sigma \in (0,1)$, depending only on n and such that, for each $x \in F$

$$\frac{m(U \cap B_t(x))}{m(B_t(x))} \ge \sigma \tag{89}$$

for some $t \in (0,\frac{1}{2}]$. Let C_1 be the constant of 4.4. We can assume $C_1 \geq 1$. Put

$$C = 2C_1\sigma^{-p} .$$

By 4.4

$$R_{k,p}(U \cup F) \leq \tfrac{1}{2}CR_{k,p}(U) . \tag{90}$$

Put

$$B = \{x;\ x \in R^n \text{ and } g(x) \neq 0\} .$$

Then $\overline{B} \subset U \cup F$ and hence $R_{k,p}(\overline{B}) \leq \frac{1}{2}CR_{k,p}(U)$, so that there exists an open set W with $\overline{B} \subset W$ and

$$R_{k,p}(W) < CR_{k,p}(U) .$$

By applying a suitable mollifier to g we can obtain a C^∞ function h with $\operatorname{spt} h \subset V \cap W$ and

$$|g-h|_{\ell,p} < \tfrac{1}{2}\varepsilon . \tag{91}$$

It follows from (88) and (91) that h has the required properties.

The main theorem can now be proved.

4.6 THEOREM

Let ℓ , m *be positive integers with* $m \leq \ell$ *and* $(\ell-m)p < n$ *and let* Ω *be a non-empty open set of* R^n . *Let* $f \in W^{\ell,p}(\Omega)$ *and be approximately continuous at every point of* Ω , *except for a set* E *with* $R_{\ell-m,p}(E) = 0$. *Let* $\varepsilon > 0$. *Then there exists a* C^m *function* g *on* Ω *such that*

(a) *the set* $F = \{x;\ x \in \Omega \text{ and } f(x) \neq g(x)\}$ *has* $R_{\ell-m,p}(F) < \varepsilon$ *and*

(b) $$|f-g|_{m,p} < \varepsilon .$$

PROOF: It can be assumed that the set $A = \{x;\ x \in \Omega \text{ and } f(x) \neq 0\}$ is not empty.

(i) We suppose to begin with that $\Omega = R^n$ and A is bounded. We show that there exists a C^m function g on R^n satisfying (a), (b) and (c) sht g is contained in the set $V = \{x;\ x \in R^n \text{ and } d(x,A) < \varepsilon\}$. Let C be the constant of 4.5. By 3.1 there exists an open set U of R^n and a C^m function h on R^n , such that $U \supset E$,

$$R_{\ell-m,p}(U) < \frac{1}{1+C}\,\varepsilon \tag{92}$$

and

$$h(x) = f(x)$$

for all $x \in R^n \sim U$. We can assume that $\operatorname{spt} h \subset V$ and $\bar{U} \subset V$. By substituting $\ell - m$ for k and $f - h$ for f in 4.5, we obtain a C^∞ function ϕ on R^n such that

$$|f-h-\phi|_{m,p} < \varepsilon\ , \tag{93}$$

$$R_{\ell-m,p}(\operatorname{spt} \phi) \le CR_{\ell-m,p}(U) \tag{94}$$

and

$$\operatorname{spt} \phi \subset V\ . \tag{95}$$

Put $g = h + \phi$. Then (b) follows from (93). Now

$$F \subset \{x;\ x \in R^n \text{ and } h(x) \neq f(x)\} \cup \operatorname{spt} \phi \subset U \cup \operatorname{spt} \phi\ , \tag{96}$$

so that by (94)

$$R_{\ell-m,p}(F) \le (1+C)R_{\ell-m,p}(U)\ .$$

It now follows from (92) that (a) is true. Since $\operatorname{spt} h$ and $\operatorname{spt} \phi$ are both contained in V, it follows that (c) is true.

(ii) We now consider the general case. The method of proof was obtained by analogy with [10]. Let $\{C_r\}_{r=0}^{\infty}$ be an infinite sequence of non-empty compact sets, such that

$$C_r \subset \operatorname{Int} C_{r+1} \tag{97}$$

for $r \ge 0$ and

$$\lim_{r\to\infty} C_r = \Omega\ . \tag{98}$$

Put $C_{-1} = \emptyset$. For each $r \ge 0$, let ϕ_r be a C^∞ function on R^n such that $0 \le \phi_r \le 1$, the set

$$M_r = \{x;\ x \in R^n \text{ and } \phi_r(x) = 1\}$$

has

$$C_r \subset \operatorname{Int} M_r \tag{99}$$

and

$$\operatorname{spt} \phi_r \subset \operatorname{Int} C_{r+1} . \tag{100}$$

Put

$$\psi_0 = \phi_0 \quad \text{and} \quad \psi_r = \phi_r - \phi_{r-1} \tag{101}$$

when $r \geq 1$. Then each ψ_r is C^∞ on R^n with compact support and

$$\operatorname{spt} \psi_r \subset (\operatorname{Int} C_{r+1}) \sim C_{r-1} . \tag{102}$$

Hence, for each $x \in \Omega$, $\psi_r(x) \neq 0$ for at most two values of r . Therefore

$$\sum_{r=0}^{\infty} \psi_r(x) = 1 \tag{103}$$

for all $x \in \Omega$. For each $r = 0,1,2,\ldots$ define

$$\left.\begin{aligned} f_r(x) &= f(x)\psi_r(x) \quad \text{when} \quad x \in \Omega , \\ &= 0 \quad \text{when} \quad x \notin \Omega . \end{aligned}\right\} \tag{104}$$

By (i), there exists for each $r \geq 0$ a C^m function g_r on R^n with compact support and such that

$$|f_r - g_r|_{m,p} < 2^{-r-1}\varepsilon , \tag{105}$$

the set

$$F_r = \{x ; x \in R^n \text{ and } f_r(x) \neq g_r(x)\}$$

has

$$R_{\ell-m,p}(F_r) < 2^{-r-1}\varepsilon \tag{106}$$

and

$$\operatorname{spt} g_r \subset (\operatorname{Int} C_{r+1}) \sim C_{r-1} . \tag{107}$$

For each $x \in \Omega$, there are at most two values of r for which $g_r(x) \neq 0$.
Hence we can define

$$g(x) = \sum_{r=0}^{\infty} g_r(x)$$

for $x \in \Omega$. It is easily seen that $F \subset \bigcup_{r=0}^{\infty} F_r$, hence

$$R_{\ell-m,p}(F) < \varepsilon .$$

Also

$$|f-g|_{m,p} \leq \sum_{r=0}^{\infty} |f_r - g_r|_{m,p} < \varepsilon .$$

BIBLIOGRAPHY

1. THOMAS BAGBY and WILLIAM P. ZIEMER, Pointwise Differentiability and Absolute Continuity, Transactions of the American Mathematical Society, 191(1974), 129-148.

2. A.P. CALDERON, Lebesgue Spaces of Differentiable Functions and Distributions, Proceedings Symposia in Pure Mathematics 4(1961), 33-49.

3. A.P. CALDERON and A. ZYGMUND, Local Properties of Solutions of Elliptic Partial Differential Equations, Studia Mathematica, 20(1961), 171-225.

4. HERBERT FEDERER and WILLIAM P. ZIEMER, The Lebesgue Set of a Function Whose Distribution Derivatives are p-th Power Summable, Indiana University Mathematics Journal, 22(1972), 139-158.

5. D. GILBARG and N.S. TRUDINGER, Elliptic Partial Differential Equations of Second Order, Springer-Verlag, 1977.

6. C. GOFFMAN, Approximation of Non-parametric Surfaces of Finite Area, Journal of Mathematics and Mechanics, 12(1963), 737-746.

7. FON-CHE LIU, A Lusin Type Property of Sobolev Functions, Indiana University Mathematics Journal, 26(1977), 645-651.

8. NORMAN G. MEYERS, A Theory of Capacities for Potentials of Functions in Lebesgue Classes, Math. Scand., 26(1970), 255-292.

9. NORMAN G. MEYERS, Taylor Expansion of Bessel Potential, Indiana University Mathematics Journal, 23(1974), 1043-1049.

10. NORMAN G. MEYERS and J. SERRIN, H = W, Proceedings of the National Academy of Sciences USA 51(1964), 1055-1056.

11. J.H. MICHAEL, The Equivalence of Two Areas for Nonparametric Discontinuous Surfaces, Illinois Journal of Mathematics, 7(1963), 59-78.

12. J.H. MICHAEL, Approximation of Functions by Means of Lipschitz Functions, Journal of the Australian Mathematical Society, 3(1963), 134-150.

13. J.H. MICHAEL, Lipschitz Approximations to Summable Functions, Acta Mathematica, III(1964), 73-95.

14. ELIAS M. STEIN, Singular Integrals and Differentiability Properties of Functions, Princeton University Press, 1970.

15. HASSLER WHITNEY, Analytic Extensions of Differentiable Functions Defined in Closed Sets, Transactions of the American Mathematical Society 36(1934), 63-89.

J.H. MICHAEL
DEPARTMENT OF MATHEMATICS
UNIVERSITY OF ADELAIDE
Box 498, G.P.O.
ADELAIDE, S.A. 5001

WILLIAM P. ZIEMER
DEPARTMENT OF MATHEMATICS
INDIANA UNIVERSITY
BLOOMINGTON, IN. 47405

Contemporary Mathematics
Volume 42, 1985

SOME PROPERTIES OF FOURIER SERIES WITH GAPS

C. J. Neugebauer

ABSTRACT. Two transformations of a function f with Fourier coefficients $\{c_n\}$ are $f \to \Sigma\, c_n e^{i\varphi(n)x}$ and $f \to \Sigma\, c_{\varphi(n)} e^{inx}$, where $\varphi(n)$ is a rearrangement of the integers. The effects of these transformations on the Hausdorff-Young inequality and the Hardy-Littlewood-Paley theorems are studied and applied.

1. This paper is concerned with various transformations on Fourier series $f \sim \sum_{-\infty}^{\infty} c_n e^{inx}$ induced by one-one maps φ from the set J of all integers into J, $\varphi: J \to J$. In particular we shall study the operators $T^{\varphi} f \sim \Sigma\, c_n e^{i\varphi(n)x}$, $T_{\varphi} f \sim \Sigma\, c_{\varphi(n)} e^{inx}$, and compositions of such operators. The notation should convey that we are looking for a function (if it exists) and its properties, e.g. $T^{\varphi} f$, whose $\varphi(n)$-th Fourier coefficient is c_n and the Fourier coefficients with index not in (J) are zero.

It turns out that the condition relevant for this purpose is the property $\varphi(J) \in \Lambda(s)$, $2 < s < \infty$, introduced by Rudin [1]. Apart from some elementary norm inequalities for the above operators, we shall study the Hausdorff-Young inequality as well as Paley's theorem on Fourier coefficients (in the trigonometric case this is due to Hardy and Littlewood [2_{II}, p.109]) under the action of φ. These results are then applied to obtain some refinements of theorem 4.5 in [4], and to a result "almost" characterizing $E \in \Lambda(s)$.

2. Let $E \subset J$ and let f be a trigonometric polynomial such that $\hat{f}(n) = 0, n \notin E$. Such polynomials are called E-polynomials in [1]. The definition of $\Lambda(s)$, $0 < s < \infty$, is due to Rudin [1].

DEFINITION. $E \in \Lambda(s)$ iff there exists $0 < r < s$ such that for every E-polynomial f, $\|f\|_s \le M \|f\|_r$.

Here and in what follows M, with or without subscripts, will be a constant not necessarily the same at different occurrences but always independent of f.

1980 Mathematics Subject Classification. 42A16.

With the use of the closed graph theorem and [1, p.225] we see that $E\in\Lambda(s)$, $2<s<\infty$, iff $\{\sum_{n\in E}|c_n|^2\}^{\frac{1}{2}} \leq M\|f\|_{s'}$, where $f\sim\sum c_n e^{inx}$ and $\frac{1}{s}+\frac{1}{s'}=1$. Since we are primarily interested in maps $\varphi:J\to J$, we say that $\varphi\in\Lambda(s)$ iff $\varphi:J\to J$ is one-one and $\varphi(J)\in\Lambda(s)$ in the above sense.

(P_0) If $\varphi\in\Lambda(s)$, $2<s<\infty$, then $\|T_\varphi f\|_2 \leq M\|f\|_{s'}$.

Proof. This is simply $\{\sum|c_{\varphi(n)}|^2\}^{\frac{1}{2}} = \|T_\varphi f\|_2$.

(P_1) If $\varphi\in\Lambda(s)$, $2<s<\infty$, then $\|T^\varphi_\varphi f\|_s \leq M\|f\|_{s'}$, where $T^\varphi_\varphi f = T^\varphi T_\varphi f\sim\sum c_{\varphi(n)}e^{i\varphi(n)x}$.

Proof. The operator T^φ_φ is, unlike T^φ and T_φ, a multiplier transformation, i.e., $\widehat{T^\varphi_\varphi f}(n) = \lambda_n\hat{f}(n)$, where $\lambda_n=0$, $n\notin\varphi(J)$, and $\lambda_n=1$, $n\in\varphi(J)$. Since $\|T^\varphi_\varphi f\|_2 = \|T_\varphi f\|_2 \leq M\|f\|_{s'}$, we obtain the desired result using standard duality arguments.

(P_2) If $\varphi\in\Lambda(s)$, $2<s<\infty$, then

$$M_1\|f\|_2 \leq \|T^\varphi f\|_\lambda \leq M_2\|f\|_2, \quad s'\leq\lambda\leq s,$$

where M_1,M_2 are also independent of λ.

Proof. Since $\|f\|_2 = \|T^\varphi f\|_2 \leq \|T^\varphi f\|_s$ and $\|T^\varphi f\|_{s'} \leq \|T^\varphi f\|_2 = \|f\|_2$, we only need to prove that $\|T^\varphi f\|_s \leq M\|T^\varphi f\|_{s'}$. Let $(T^\varphi f)_N = \sum_{|\varphi(n)|\leq N} c_n e^{i\varphi(n)x}$, the N-th partial sum. Then, for some $g_{s'}=1$, we have by Parseval's formula, and P_1

$$\begin{aligned}\|(T^\varphi f)_N\|_s &= \int \overline{(T^\varphi f)_N}\cdot g = \int \overline{(T^\varphi f)_N}\cdot T^\varphi_\varphi g \\ &\leq \|(T^\varphi f)_N\|_{s'}\|T^\varphi_\varphi g\|_s \leq M\|(T^\varphi f)_N\|_{s'},\end{aligned}$$

and letting $N\to\infty$ the proof is complete.

(P_3) If $\psi\in\Lambda(s_1)$, $\varphi\in\Lambda(s_2)$, $2<s_1,s_2<\infty$, then $\|T^\psi_\varphi f\|_{s_1} \leq M\|f\|_{s_2'}$, where $T^\psi_\varphi = T^\psi T_\varphi$.

Proof. By P_2, $\|T^\psi_\varphi f\|_{s_1} \leq M_1\|T_\varphi f\|_2 \leq M_2\|f\|_{s_2'}$.

3. In this section we will study the action of $\varphi\in\Lambda(s)$ upon the Hausdorff-Young inequality: $\|c_n\|_{p'} \leq M\|f\|_p$ and $\|f\|_{p'} \leq M\|c_n\|_p$, $1<p\leq 2$, where $\|c_n\|_r = \{\sum|c_n|^r\}^{1/r}$ ([2_{II}, p.101]).

THEOREM 1. Let $\varphi\in\Lambda(s)$, $2<s<\infty$, and let $1<p\leq 2$. Then

(i) $\|c_{\varphi(n)}\|_{p'} \leq M\|f\|_\lambda$, $\lambda = \frac{sp'}{sp'-2}$

(ii) $\|T^\varphi f\|_{\frac{sp'}{2}} \leq M\|c_n\|_p$.

Proof. (i) The proof is accomplished by applying the Riesz-Thorin interpolation theorem [2_{II}, p.95] to $\|c_{\varphi(n)}\|_2 \leq M\|f\|_{s'}$ and $\|c_{\varphi(n)}\|_\infty \leq M\|f\|_1$.

(ii) The proof follows standard lines (see e.g. [2_{II}, p.103]). Let

$(T^{\varphi}f)_N = \sum_{|\varphi(n)|\leq N} c_n e^{i\varphi(n)x}$. For some $\|g\|_\lambda = 1$, $g \sim \sum c_n e^{inx}$, we have

$$\|(T^{\varphi}f)_N\|_{\frac{sp'}{2}} = \int \overline{(T^{\varphi}f)_N}\, g = \sum \overline{c}_j d_{\varphi(j)}$$

$$\leq \|c_j\|_p \|d_{\varphi(j)}\|_{p'} \leq M\|c_j\|_p \quad \text{by (i).}$$

The result now follows by letting $N\to\infty$.

4. This section is similar to the preceding one in that we will investigate the Hardy-Littlewood-Paley theorem on Fourier coefficients [2_{II}, pp.109-120] subject to $\varphi\varepsilon\Lambda(s)$.

THEOREM 2. Let $\varphi\in\Lambda(s)$, $2<s<\infty$, and let $1<p\leq 2$. Then

(i) $\{\sum |c_{\varphi(j)}|^p(|j|+1)^{p-2}\}^{1/p} \leq M\|f\|_\lambda$, $\lambda = \frac{sp'}{sp'-2}$

(ii) $\|T^{\varphi}f\|_{sp'/2} \leq M\{\sum |c_j|^{p'}(|j|+1)^{p'-2}\}^{1/p'}$.

Proof. (i) We map $f\sim\sum c_n e^{inx}$ into the sequence $\{Tf(j)\} = \{c_{\varphi(j)}(|j|+1)\}$. If μ is the measure that assigns $\frac{1}{(|j|+1)^2}$ to j and 0 to any set not containing an integer, we get, since $\varphi(J)\subset\Lambda(s)$,

(1) $\|Tf\|_{2,\mu} = \{\sum |c_{\varphi(j)}|^2\}^{\frac{1}{2}} = \|T_{\varphi}f\|_2 \leq M\|f\|_{s'}$.

Using exactly the same argument as in [2_{II}, p.121] we can show that T is of weak type (1,1), i.e.,

(2) $$\mu\{|Tf(j)| > y\} \leq \frac{M\|f\|_1}{y} .$$

The Marcinkiewicz interpolation theorem now yields (i).

(ii) With the same notation as in section 3 we have

$$\|(T^{\varphi}f)_N\|_{sp'/2} = \int \overline{(T^{\varphi}f)_N}\cdot g = \sum \overline{c}_j d_{\varphi(j)} =$$

$$\sum \overline{c}_j(|j|+1)^{1-2/p'} d_{\varphi(j)}(|j|+1)^{2/p'-1}$$

$$\leq \{\sum |c_j|^{p'}(|j|+1)^{p'-2}\}^{1/p'}\{\sum |d_{\varphi(j)}|^p(|j|+1)^{p-2}\}^{1/p}$$

$$\leq M\{\sum |c_j|^{p'}(|j|+1)^{p'-2}\}^{1/p'} \quad \text{by (i) and } \|g\|_\lambda = 1 .$$

5. With the use of the preceding results we obtain in this section a refinement of Rudin's theorem 4.5 [1]. If $E=\{n_1<n_2<\dots\}$ is a subset of the positive integers and t is a positive integer, we define $r_t(n,E)$ by $\sum_1^\infty r_t(n,E)z^n = (\sum z^{n_k})^t$, i.e., $r_t(n,E)$ is the number of representations $n=n_{i_1}+\dots+n_{i_t}$, $n_{i_j}\in E$, with permutations counted as different representations.

THEOREM 3. Let $E\in\Lambda(s)$, $E=\{n_1<n_2<\dots\}$, be a subset of the positive integers, let $t\geq 2$ be an integer, and let $2\leq q<\infty$.

(i) $tq' \le s$ implies $\{\sum_1^N r_t(n,E)^q\}^{1/q} = O(N^{t/s})$.

(ii) $tq' \ge s$ implies $\{\frac{1}{N}\sum_1^N r_t(n,E)q\}^{1/q} = O(N^{2t/s^{-1}})$.

REMARK. We point out that in our case s need not be an integer. If, however, $s=2t$ and $q=2$, the above theorem reduces to Rudin's result [1, p.218].

Proof. Let $f = \sum_{j=1}^k e^{in_jx}$. By the Hausdorff-Young inequality

$$\{\sum_1^{n_k} r_t(n,E)^q\}^{1/q} \le M\{\int |f|^{tq'}\}^{\frac{t}{tq'}} = M\{\int |T f_0|^{tq'}\}^{\frac{t}{tq'}},$$

where $f_0 = \sum_1^k e^{inx}$ and $\varphi : J \to \varphi(J) = E$ is such that $\varphi(j) = n_j$, $1 \le j \le k$.

(i) $tq' \le s$. Since $\Lambda(s) \subset \Lambda(tq')$ we can apply theorem 2 (with $p=2=p'$ and $s=tq'$) and obtain

$$\{\int |T^{\varphi} f_0|^{tq'}\}^{t/tq'} \le M\{\sum_1^k 1\}^{t/2} = Mk^{t/2}.$$

The rest of the proof of (i) is now the same as in [1, p.218], and we give it here for the sake of completeness. By [1, p.213], $ck^{s/2} \le n_k$, where c does not depend on k. If we choose $c(k-1)^{s/2} \le N < ck^{s/2}$, then

$$\{\sum_1^N r_t(n,E)^q\}^{1/q} \le \{\sum_1^{n_k} r_t(n,E)^q\}^{1/q} \le M_1 k^{t/2} \le M_2 N^{t/s}.$$

(ii) $tq' \ge s$. We apply now theorem 2 with $p' = \frac{2tq'}{s}$ and obtain

$$\{\sum_1^{n_k} r_t(n,E)^q\}^{1/q} \le M\{\sum_1^k n^{p'-2}\}^{t/p'} \le Mk^{t/p}.$$

As in (i),

$$\{\sum_1^N r_t(n,E)^q\}^{1/q} \le M \cdot N^{2t/sp} = M \cdot N^{\frac{2t}{s} - \frac{1}{q'}}.$$

If we divide both sides by $N^{1/q}$ the proof of (ii) is complete.

There is a version of theorem 3 for the case $1<q\le 2$.

THEOREM 4. Under the same hypothesis as in theorem 3, except that $1<q\le 2$, we have

(i) $tq \le s$ implies $\{\sum_1^N r_t(n,E)^q n^{q-2}\}^{1/q} = O(N^{t/s})$

(ii) $tq \ge s$ implies $\{N \sum_1^N r_t(n,E)^q n^{q-2}\}^{1/q} = O(N^{2t/s})$.

Proof. With the same f as in theorem 3 we obtain using Paley's theorem [2_{II}, p.121]

$$\{\sum_1^{n_k} r_t(n,E)^q n^{q-2}\}^{1/q} \le M\{\int |f|^{tq}\}^{t/tq}.$$

The two cases are handled in precisely the same way as in theorem 3.

6. Let us call a (not necessarily linear) map $T: B^2 \to \ell^p$ a B^2-bounded map iff $\|T\sigma\|_p \le M\|\sigma\|_2$, $\sigma\in B^2$, where B^2 is the unit ball in ℓ^2. This notion will be useful in "almost" characterizing $\Lambda(s)$, and, in case s is an even integer, will in fact provide a characterization.

Let $E=\{n_1<n_2<\dots\}$ be a subset of the positive integers, and let j be a positive integer. Define a map $T=T(E,j)$ on B^2 by $T\{\gamma_k\} = \{\Gamma(n,j)$, where $\Gamma(n,j)$ is given by the formula

$$(\sum_k \gamma_k z^{n_k})^j = \sum_n \Gamma(n,j)z^n .$$

Let $2<s<\infty$, and let $1\le j_1\le j_2$ be integers such that $1 < \frac{s}{j_2} \le 2 \le \frac{s}{j_1}$. We set $\sigma_i=s/j_i$ and consider the maps $T_i=T(E,j_i)$, $i=1,2$.

THEOREM 5.

(i) If $E\in\Lambda(s)$, then $T_2:B^2 \to \ell^{\sigma_2'}$ is B^2-bounded.

(ii) If $T_1: B^2 \to \ell^{\sigma_1'}$ is B^2-bounded, then $E\in\Lambda(s)$.

Proof. (i) Let $\{\gamma_k\}\in B^2$, and let $f\sim\sum \gamma_k e^{in_kx}$. By the Hausdorff-Young theorem we have

$$\|T_2\{\gamma_k\}\|_{\sigma_2'} = \{\sum|\Gamma(n,j_2)|^{\sigma_2'}\}^{1/\sigma_2'} \le$$

$$M\{\int|f|^s\}^{j_2/s} \le M\|\{\gamma_k\}\|_2^{j_2} \le M\|\{\gamma_k\}\|_2 ,$$

where the last inequality is due to $\{\gamma_k\}\in B^2$, and the next to the last inequality follows from theorem 2 with $p'=2$.

(ii) We have to show that $\{\sum|c_{n_k}|^2\}^{\frac12} \le M\|f\|_{s'}$, $f\sim\sum c_n e^{inx}$. If $\sum|\gamma_{n_k}|^2=1$, then (all sums extend over a finite index set)

$$\sum_{k=1}^{K} c_{n_k}\gamma_{n_k} = \int f(t)\sum\gamma_{n_k} e^{-in_kt}dt$$

$$\le \|f\|_{s'}\|\sum \gamma_{n_k} e^{-in_kt}\|_s = \|f\|_{s'}\{|\sum \Gamma(n,j_1)e^{-int}|^{\sigma_1}dt\}^{1/s}$$

$$\le M\|f\|_{s'}\{\sum|\Gamma(n,j_1)|^{\sigma_1'}\}^{\frac{1}{j_1\sigma_1'}}$$

$$\le M\|f\|_{s'}\|\{\gamma_{n_k}\}\|_2^{1/j_1} \le M\|f\|_{s'}$$

and the proof is complete by letting $K\to\infty$.

REMARK. If s is an even integer >2, we can choose $j_1=j_2$, $\sigma_1=\sigma_2=2$. We then have a characterization of $E\in\Lambda(s)$ in terms of the B^2-boundedness of the map $T_1=T_2:B^2 \to \ell^2$.

BIBLIOGRAPHY

1. W. Rudin, "Trigonometric series with gaps", J. Math. Mech., 9 (1960), 202-227.

2. A. Zygmund, Trigonometric series, Cambridge University Press, Cambridge, 1959.

DEPARTMENT OF MATHEMATICS
PURDUE UNIVERSITY
WEST LAFAYETTE, INDIANA 47907

Contemporary Mathematics
Volume **42**, 1985

AN EXTENSION OF THUNSDORFF'S INTEGRAL INEQUALITY TO A CLASS OF MONOTONE FUNCTIONS

Togo Nishiura

A well known inequality in analysis is

$$\left(\int_0^1 f^m dx\right)^{1/m} \leq \left(\int_0^1 f^n dx\right)^{1/n} \tag{1}$$

where f is a nonnegative measurable function on $[0,1]$ and $0 < m \leq n < \infty$. In his thesis [1], H. Thunsdorff proved the following theorem.

THEOREM 1. Let $0 < m \leq n < \infty$ and f be a nonnegative convex function on $[0,1]$ satisfying $f(0) = 0$. Then

$$\left((m+1)\int_0^1 f^m dx\right)^{1/m} \leq \left((n+1)\int_0^1 f^n dx\right)^{1/n}. \tag{2}$$

The theorem is quoted in [2], page 307, without proof. F. Schnitzer and the author gave an elementary proof of Thunsdorff's theorem in [3]. The present note investigates the inequality (2) when convexity is replaced by monotonicity.

In general one can prove, with the aid of the classical inequality (1), the following.

THEOREM 2. Let $0 < m \leq n < \infty$ and f be a nonegative measurable function on $[0,1]$. Then

$$\left((m+1)\int_0^1 f^m dx\right)^{1/m} \leq e\left((n+1)\int_0^1 F^n dx\right)^{1/n}. \tag{3}$$

Moreover, the constant e is sharp as the monotone function $f = 1$ shows.

In order to state the generalization of Thunsdorff's theorem, we need some notation. For a real valued function f on $[0,1]$, we define two subsets of $\mathbf{R}^2$ as follows.

$$A^*(f) = \text{closure of } \{(x,y) \mid y \geq f(x) \text{ and } x \in [0,1]\},$$
$$A_*(f) = \text{closure of } \{(x,y) \mid y \geq f(x) \text{ and } x \in [0,1]\}.$$

1980 Mathematics Subject Classification. 26D15

A classical concept is that of a starlike set. A set $M \subset \mathbf{R}^2$ is said to be <u>starlike with respect to the origin of $\mathbf{R}^2$</u> if the origin is a member of M and $L \cap M$ is connected for each line L in $\mathbf{R}^2$ passing through the origin. Clearly, if f is a convex function on $[0,1]$ with $f(0) = 0$ then $A^*(f)$ is starlike with respect to the origin. We state our main theorem.

THEOREM 3. Let $0 < m \le n < \infty$ and f be a nonnegative nondecreasing function on $[0,1]$ with $A^*(f)$ starlike with respect to the origin. Then inequality (2) is true.

To prove this theorem we need a geometric lemma which was observed in [3].

LEMMA. Let F and G be two nonnegative nondecreasing functions on $[0,1]$ satisfying the following conditions: There is $p \in [0,1]$ such that

(i) $F(x) \ge G(x)$ for $0 \le x < p$;
(ii) $F(x) \le G(x)$ for $p < x \le 1$;
(iii) The above regions have equal areas, i.e.

$$\int_0^p (F-G)dx = \int_p^1 (G-F)dx .$$

Then, for $k \ge 0$,

$$\int_0^1 F^{k+1}dx \le \int_0^1 G^{k+1}dx . \tag{4}$$

Moreover, equality holds exactly when $F = G$ almost everywhere.

PROOF. It is easily seen that

$$\int_0^p \int_{G(x)}^{F(x)} y^i dydx \le \int_p^1 \int_{F(x)}^{G(x)} y^k dydx .$$

Hence

$$0 \le \int_0^1 \int_{F(x)}^{G(x)} y^k dydx ,$$

which in turn gives

$$\int_0^1 \int_0^{F(x)} y^k dydx \le \int_0^1 \int_0^{G(x)} y^k dydx .$$

The Lemma now follows.

PROOF OF THEOREM 3. It is given that f is nonnegative, nondecreasing, $f(0) = 0$ and $A^*(f)$ is starlike with respect to the origin. Hence, for each line $y = Cx$, either $Cx \ge f(x)$ for all x or there is $p \in [0,1]$ such that $Cx \ge f(x)$ for $0 \le x < p$ and $Cx \le f(x)$ for $p < x \le 1$.

Suppose $0 < m \le n < \infty$. Choose C_m so that $F(x) = (C_m x)^m$ and

$G(x) = f^m(x)$ satisfies (i), (ii) and (iii) of the Lemma above. One easily obtains

$$C_m = ((m+1)\int_0^1 f^m dx)^{1/m} .$$

Next choose $k = (n-m)/m$. Then (4) gives

$$C_m^n \int_0^1 x^n dx \le \int_0^1 f^n dx .$$

Theorem 3 now follows. This proof is exactly the same as the one given in 3 where f is convex.

The geometric lemma above also yields the following inequality.

THEOREM 4. Let $0 < m \le n < \infty$ and f be a nonnegative nondecreasing function on $[0,1]$ with $A_*(f)$ starlike with respect to the origin. Then

$$((m+1)\int_0^1 f^m dx)^{1/m} \ge ((n+1)\int_0^1 f^n dx)^{1/n} .$$

PROOF. In this proof let $G(x) = (C_m x)^m$ and $F(x) = f^m(x)$ be such that (i), (ii) and (iii) of the Lemma are satisfied. Then (4) yields

$$\int_0^1 f^n dx \le C_m^n \int_0^1 x^n dx ,$$

with $C_m = ((m+1)\int_0^1 f^m dx)^{1/m}$, and the theorem is proved.

Finally, we give examples related to Theorems 3 and 4. We will need the following proposition in our verifications of the examples. The proof of the proposition is elementary.

PROPOSITION. Let f be a nonnegative measureable function on $[0,1]$. Then

$$\lim_{n\to\infty} ((n+1)\int_0^1 f^n dx)^{1/n} = \operatorname{ess\,sup} f .$$

EXAMPLE 1. There is a nonnegative nondecreasing function G on $[0,1]$ and three numbers $0 < u < v < w < \infty$ such that

$$((u+1)\int_0^1 G^u dx)^{1/u} < ((v+1)\int_0^1 G^v dx)^{1/v} ,$$

$$((w+1)\int_0^1 G^w dx)^{1/w} < ((v+1)\int_0^1 G^v dx)^{1/v} .$$

PROOF. In the geometric lemma let $F(x) = x$ for $0 \le x \le 1$ and G be such that $F - G$ is not Lebesgue equivalent to 0 and $\lim_{x\to 1} G(x) = 1$. Then for $k+1 = n>1$ we have

$$\frac{1}{n+1} = \int_0^1 x^n dx < \int_0^1 G^n dx .$$

Since

$$\int_0^1 x\,dx = \int_0^1 G\,dx ,$$

$$2\int_0^1 G\,dx < ((n+1)\int_0^1 G^n dx)^{1/n} .$$

The proposition completes the proof.

EXAMPLE 2. There is a nonnegative nondecreasing function F on $[0,1]$ and three numbers $0 < u < v < w < \infty$ such that

$$((u+1)\int_0^1 F^u dx)^{1/u} > ((v+1)\int_0^1 F^v dx)^{1/v} ,$$

$$((w+1)\int_0^1 F^w dx)^{1/w} > ((v+1)\int_0^1 F^v dx)^{1/v} .$$

PROOF. In the geometric lemma let $G(x) = x$ for $0 \le x \le 1$ and F be such that $F - G$ is not Lebesgue equivalent to 0 and $\lim_{x \to 1} F(x) = 1$. The proof is now easily completed.

BIBLIOGRAPHY

1. H. Thunsdorff, "Konvexe Funktionen und Ungleichungen", Inaugural-Dissertation, Göttingen 1932.

2. D. S. Mitrinović, Analytic Inequalities, Springer-Verlag, Berlin, 1970.

3. T. Nishiura and F. Schnitzer, "A proof of an inequality of H. Thunsdorff, Publications de la faculté d'électrotechnique de l'université à Belgrade", Série: Mathématique et Physique, No. 357-380 (1971), pages 1-2.

DEPARTMENT OF MATHEMATICS
WAYNE STATE UNIVERSITY
DETROIT, MICHIGAN 48202

Contemporary Mathematics
Volume 42, 1985

ON GENERALIZED BOUNDED VARIATION

L. Di Piazza and C. Maniscalco[1]

ABSTRACT. We introduce some classes [G Λ BV] of functions of generalized Λ-bounded variation, and [GΦ-BV] of functions of generalized Φ-bounded variation. It is shown that the union, taken over all sequences Λ, of the [G Λ BV] as well as the union, taken over all functions Φ, of the [GΦ-BV] is the class of all continuous a.e. functions.

Many authors generalized the notion of bounded variation; in most cases these new concepts were introduced by investigating the behavior of Fourier series. In 1972 in [8] Waterman defined the classes of functions of Λ-bounded variation, whose genesis arises in the work of Goffman and Waterman about the everywhere convergence of Fourier series and the everywhere convergence of Fourier series for every change of variable [4], [5], [6]. In [8] and [9] Waterman proves various results on the summability and convergence of Fourier series of functions of ΛBV. In 1978 in [2] Bongiorno and Vetro gave the notion of variation with regard to a complete subbase $\mathcal{A}$, and proved that a theorem of Riesz's type holds for such functions.

In this note we introduce the concept of generalized Λ-bounded variation (G ΛBV) that is a generalization of both the Waterman and the Bongiorno-Vetro variations. We prove some properties of the classes [G Λ BV]; in particular we show that the union, taken over all sequences Λ, of the [G Λ BV] is the class of all continuous a.e. functions.

Moreover we generalize the concept of function of Φ-bounded variation in the sense of L. C. Young introducing the classes of functions [GΦ-BV]. We also prove that the union of the classes

---

1980 Mathematics Subject Classification. 26A45.
[1]This research was supported by the Ministero della Pubblica Instruzione (Italy).

$[G\Phi\text{-BV}]$, taken over all functions Φ, is the class of all continuous a.e. functions.

Finally we observe that the Fourier series of functions of classes defined here are convergent a.e.

1. Let $\mathcal{A}$ be a family of intervals in $[a,b]$ such that for almost every $x \in [a,b]$ there is a constant $\delta(x) > 0$ for which the intervals $[x-\gamma,x]$, $[x,x+\gamma]$ are in $\mathcal{A}$, if $0 < \gamma < \delta(x)$. $\mathcal{A}$ is called a <u>complete</u> <u>subbase</u> (c.s.). The domain of δ is denoted $\mathcal{D}(\mathcal{A})$.

Let us suppose that f is a real function on an interval $[a,b]$; let Λ denote a non-decreasing sequence $\{\lambda_i\}$ of positive real numbers such that $\Sigma_i 1/\lambda_i = \infty$; Let $\{I_i\}$ be a sequence of non-overlapping intervals $I_i = [a_i,b_i]$ in a complete subbase $\mathcal{A}$ and write $f(I_i) = f(b_i)-f(a_i)$; let $V_\Lambda(f,\mathcal{A})$ denote the supremum of $\Sigma_i |f(I_i)|/\lambda_i$ over all sequences $\{I_i\}$ of non-overlapping intervals of $\mathcal{A}$. Throughout this note, when we consider a collection of intervals, they will always be non-overlapping.

DEFINITION. A function f is said to be of <u>generalized</u> <u>Λ-bounded</u> <u>variation</u> ($G\Lambda BV$) if $V_\Lambda(f,\mathcal{A}) < \infty$, for some c.s. $\mathcal{A}$.

It may be shown readily that the following properties are equivalent to the definition of $G\Lambda BV$:

i) There is a complete subbase $\mathcal{A}$ such that, for every sequence $\{I_i\}$ of intervals in $\mathcal{A}$, $\Sigma_i |f(I_i)|/\lambda_i < \infty$.

ii) There is a complete subbase $\mathcal{A}$ and a real $M > 0$ such that, for every finite collection $\{I_i\}$ $(i = 1,2,\dots,N)$ of intervals in $\mathcal{A}$, $\sum_i^N |f(I_i)|/\lambda_i < M$.

Let $[G\Lambda BV]$ denote the class of all functions of $G\Lambda BV$; in general, let $[P]$ denote the class of all functions with property P. If $\Lambda = \{i\}$, we say that f is of <u>generalized</u> <u>harmonic</u> <u>bounded</u> <u>variation</u> (GHBV) and we write $V_\Lambda(f,\mathcal{A}) = V_H(f,\mathcal{A})$.

Let $V(f,\mathcal{A})$ denote the supremum of $\Sigma_i |f(I_i)|$ over all sequences $\{I_i\}$ of intervals of $\mathcal{A}$ [2]. If $V(f,\mathcal{A}) < \infty$ for some c.s. $\mathcal{A}$, we say that f is GBV.

Then we have:

THEOREM 1. For each complete subbase $\mathcal{A}$

$$V(f,\mathcal{A}) = \sup_\Lambda V_\Lambda(f,\mathcal{A}) .$$

Moreover $V(f,\mathcal{A}) < \infty$ if and only if $V_\Lambda(f,\mathcal{A}) < \infty$ for every Λ.

PROOF. To prove the first part, it is enough to show that $V(f,\mathcal{A}) \le \sup_\Lambda V_\Lambda(f,\mathcal{A})$. We first suppose $V(f,\mathcal{A}) < \infty$.

Given $\varepsilon > 0$, let $\{I_i\}$ be a sequence of intervals of $\mathcal{A}$ such that $V(f,\mathcal{A}) - \varepsilon/2 < \sum_1^\infty |f(I_i)|$.

Let $m \in \mathbb{N}$ and $\beta \in]0,1[$ be such that $\sum_{m}^{\infty} |f(I_i)| < \epsilon/4$ and $(1-1/m^{\beta}) < \epsilon/\left(4 \sum_{1}^{\infty} |f(I_i)|\right)$. Then

$$\sum_{1}^{\infty} |f(I_i)|(1-1/i^{\beta}) \leq \sum_{m}^{\infty} |f(I_i)| + (1-1/m^{\beta}) \sum_{1}^{m-1} |f(I_i)| < \epsilon/2 .$$

Denote by Λ_ϵ the sequence $\{i^{\beta}\}$; we have

$$V(f,\mathcal{A}) \leq \epsilon + V_{\Lambda_\epsilon}(f,\mathcal{A}) \leq \epsilon + \sup_{\Lambda} V_{\Lambda}(f,\mathcal{A})$$

and the conclusion follows from the arbitrariness of $\epsilon > 0$.

Now let us suppose that $V(f,\mathcal{A}) = \infty$. If $\sup_{\Lambda} V_{\Lambda}(f,\mathcal{A}) = S < \infty$, there is a sequence $\{I_i\}$ of intervals of $\mathcal{A}$ such that $S + 1 < \sum_{i} |f(I_i)| < \infty$.

Then if $0 < \epsilon < 1$ there is a real β, $0 < \beta < 1$, such that $\sum_{i} |f(I_i)|(1-1/i^{\beta}) < \epsilon$, from which it follows that

$$S + 1 < \epsilon + \sum_{i} 1/i^{\beta} |f(I_i)| < \epsilon + S ,$$

and this is a contradiction.

To prove the second part of the theorem, it is possible to proceed as in [7] theorem 5, after observing that, for every interval I of $\mathcal{A}$, $|f(I)| \leq B$, where $B = \inf_{\Lambda=\{\lambda_1,\lambda_2,\ldots\}} \{V_{\Lambda}(f,\mathcal{A}) \cdot \lambda_1\}$. ■

Theorem 1 cannot be improved in the sense that if we have an infinite set of sequences $\Lambda_n = \{\lambda(n,k)\}_{k \in \mathbb{N}}$, then there is a function in $\bigcap_n [G\Lambda_n BV]$ that is not GBV, as the following example shows.

EXAMPLE. For each positive integer n, let $\Lambda_n = \{\lambda(n,k)\}$ be a divergent non-decreasing sequence of positive numbers such that $\sum_{k=1}^{\infty} 1/\lambda(n,k) = \infty$, and let $\Lambda^n = \{1/\ell(n,k)\}$ be the sequence with $\ell(n,k) = \sum_{i=1}^{n} 1/\lambda(i,k)$. There exists ([7] lemma 2) a decreasing sequence $\{B(k)\}$ of positive numbers tending to zero, such that $\sum_{1}^{\infty} B(k) = \infty$ and $\sum_{k=1}^{\infty} B(k)\ell(n,k) < \infty$ $(n = 1,2,\ldots)$. Fix a complete subbase $\mathcal{A}'$ in $[a,b]$, and choose a sequence $\{I_i\} = \{[a_i,b_i]\}$ of intervals in $\mathcal{A}'$ such that $a \leq a_1 < b_1 \leq \ldots \leq a_n < b_n \leq \ldots < b$ (this choice is always possible). We define the function f as follows: $f(a_i) = \sum_{1}^{i} (-1)^k B(k)$, $f(b_i) = \sum_{1}^{i+1} (-1)^k B(k)$, $f(a) = 0$ if $a < a_i$, $f(b) = \sum_{1}^{\infty}(-1)^k B(k)$ and linear otherwise. Therefore

$$V_{\Lambda_n}(f,\mathcal{A}') \leq V_{\Lambda^n}(f,\mathcal{A}') = \sum_{k=1}^{\infty} B(k)\ell(n,k) < \infty \quad (n = 1,2,\ldots).$$

Let $\mathcal{A}$ be a generic complete subbase; since $\mathcal{A}$ is a Vitali's covering of $I_i \cap \mathcal{D}(\mathcal{A})$ $(i = 1,2,\ldots)$, for each positive integer i

there exists a finite or denumerable family of intervals $\{J_k^i\} = \{[\alpha_k^i, \beta_k^i]\}$ such that $\mu([I_i \cap \mathcal{B}(\mathcal{A})] - \bigcup_k J_k^i) = 0$, in denoting Lebesgue measure. Because f is linear in I_i, $\sum_k |f(J_k^i)| = |f(I_i)|$. Therefore

$$V(f,\mathcal{A}) = \sum_i \sum_k |f(J_k^i)| = \sum_i B(i) = \infty .$$

Our main result is the following.

THEOREM 2. The function f is of class $[G \Lambda BV]$ on $[a,b]$ for some sequence Λ if and only if it is continuous almost everywhere in $[a,b]$.

PROOF. If $V_\Lambda(f,\mathcal{A}) < \infty$ for some c.s.$\mathcal{A}$, for each positive integer k the set $E_k = \{x \in \mathcal{L}(\mathcal{A}) : \omega(f,x) > 1/k\}$, where $\omega(f,x)$ is the oscillation of f at x, is finite. In fact, if E_k is infinite, in E_k there exists a monotone sequence $\{x_i\}$. Let I_i be an interval in $\mathcal{A}$ with x_i as an endpoint, length smaller than $1/2 \inf_{m \neq i} d(x_i,x_m)$, and such that $|f(I_i)| \geq 1/(2k)$. Intervals $\{I_i\}$ are non-overlapping and such that $\sum_1^\infty 1/\lambda_i |f(I_i)| \geq 1/(2k) \sum_1^\infty 1/\lambda_i = \infty$; so f cannot be in $[G \Lambda BV]$. Therefore f is continuous a.e.

Conversely, if f is continuous a.e. in $[a,b]$, let E be the set of all points of continuity for f. Let $\{C_m\}$ be a monotonically increasing sequence of closed sets in E, such that $\mu(C_m) \geq b-a-1/m$ and $\mu(\bigcup_m C_m) = \mu(E)$. We define a function $\delta(x)$ in $\bigcup_m C_m$ so that the oscillations $\omega(f,[x-\delta(x),x])$ and $\omega(f,[x,x+\delta(x)])$ are smaller than $1/m$ if x is in $C_m - C_{m-1}$.

Let $\mathcal{A}$ be the complete subbase consisting precisely of all the intervals $[x-\gamma,x]$, $[x,x+\gamma]$ if $x \in \bigcup_m C_m$ and $0 < \gamma < \delta(x)$.

For $\rho > 0$ define

$$\omega(f,\mathcal{A},\rho) = \sup\{|f(x')-f(x'')| : |x'-x''| < \rho \text{ and } [x',x''] \in \mathcal{A}\}.$$

This is an increasing function of ρ, and $\lim_{\rho\to 0} \omega(f,\mathcal{A},\rho) = 0$.

In fact if $\lim_{\rho\to 0} \omega(f,\mathcal{A},\rho) = \ell > 0$, for every sequence $\{\rho_i\} \searrow 0$ there exists a sequence of intervals $[x_i',x_i''] \in \mathcal{A}$ so that $|x_i'-x_i''| < \rho_i$ and $|f(x_i')-f(x_i'')| > \ell/2$. By the construction of $\mathcal{A}$, these intervals $[x_i',x_i'']$ have at least one endpoint in $C_{[2/\ell]}$; we indicate such an endpoint by y_i. We may suppose that $\{y_i\}$ is convergent to a point $y \in C_{[2/\ell]}$. Then, because f is continuous in y, we have a contradiction.

Let $\{I_i\}$ be a sequence of intervals in $\mathcal{A}$. For each positive integer m we define

$T_m = \{I_h : \omega(f,\mathcal{A},(b-a)/m) \geq |f(I_h)| > \omega(f,\mathcal{A},(b-a)/(m+1))\}$.

For every $I_h \in T_m$, $\mu(I_h) > (b-a)/(m+1)$; thus the intervals may be relabeled J_k so that

$$|f(J_1)| \geq |f(J_2)| \geq \ldots \geq |f(J_i)| \geq \ldots \to 0 .$$

Then we have $|f(J_i)| \leq \omega(f,\mathcal{A},(b-a)/i)$. Let $\Lambda = \{\lambda_i\}$ be a sequence such that $\lambda_i \uparrow \infty$, $\sum_1^\infty 1/\lambda_i = \infty$ and $\sum_1^\infty 1/\lambda_i\, \omega(f,\mathcal{A},(b-a)/i) < \infty$ ([7] lemma 1). So we have $\sum_1^\infty |f(I_i)|/\lambda_i \leq \sum_1^\infty |f(J_i)|/\lambda_i < \infty$, and so $f \in [G\Lambda BV]$. ∎

2. Suppose that $\Phi(x) \geq 0$ is a convex function such that $\Phi(x) = o(x)$ as $x \to 0^+$, $\Phi(x)/x \to +\infty$ as $x \to +\infty$, and $\Phi(0) = 0$.

DEFINITION. A function f on $[a,b]$ is said to be of <u>generalized</u> Φ-<u>bounded</u> <u>variation</u> (GΦ-BV) if there exists a complete subbase $\mathcal{A}$ and a constant $M > 0$ such that, for every sequence $\{I_i\}$ of intervals of $\mathcal{A}$, we have $\sum_i \Phi(|f(I_i)|) < M$.

THEOREM 3. The union, taken over all functions Φ, of the [GΦ-BV] on $[a,b]$ is the class of continuous almost everywhere functions on $[a,b]$.

PROOF. It is enough to show that if f is continuous a.e. on $[a,b]$, f is GΦ-BV for some Φ. Let $\mathcal{A}$ be the complete subbase constructed in Theorem 2. Let $\{n_i\}$ be the sequence of natural numbers with the following properties: n_1 is the greatest integer such that $\omega(f,\mathcal{A},b-a) = \omega(f,\mathcal{A},(b-a)/n_1) = \xi_1$, n_2 is the greatest integer such that $\omega(f,\mathcal{A},(b-a)/(n_1+1)) = \omega(f,\mathcal{A},(b-a)/(n_1+n_2)) = \xi_2$ and, at the ith step, n_i is the greatest integer such that $\omega(f,\mathcal{A},(b-a)/(n_1+n_2+\ldots+n_{i-1}+1)) = \omega(f,\mathcal{A},(b-a)/(n_1+n_2+\ldots+n_i)) = \xi_i$.

Let $\varphi(t)$ be a continuous function defined in $[0,+\infty)$ such that $\varphi(0) = 0$, $\lim_{x\to+\infty} \varphi(x) = +\infty$, $\varphi(\xi_i) = 1/\bar{n}_i\; 1/(n_1+n_2+\ldots+n_i)^2$ where $\bar{n}_i = \max(n_1,n_2,\ldots,n_i)$, and linear otherwise.

Proceeding as in Theorem 2, we see that the function f is GΦ-BV, with $\Phi(x) = \int_0^x \varphi(t)dt$. ∎

3. Let f be a function of period 2π and summable in $[-\pi,\pi]$.

We observe that if f is $G\Lambda BV$ for some Λ then the Fourier series of f converges to $f(x)$ a.e. In fact, if N is the set of all points of unboundness for f then $A = N'$ is an open set and, so, $A = \bigcup_i\,]a_i,b_i[$. For each real positive ϵ, let $C_\epsilon = \bigcup_i [a_i + \epsilon/2^{i+1}, b_i - \epsilon/2^{i+1}]$; let us define the function

$f_\varepsilon(x) = f(x)$ if $x \in C_\varepsilon$ and $f_\varepsilon(x) = 0$ if $x \notin C_\varepsilon$, whose Fourier series is convergent to $f_\varepsilon(x)$ a.e. [3]. Then, by the localization principle, the Fourier series of f converges to $f(x)$ for almost every $x \in C_\varepsilon$. By the arbitrariness of ε we obtain the convergence of the Fourier series of f to $f(x)$ a.e.

In the Λ-bounded variation case Waterman proved in [8] that if f is HBV its Fourier series converges everywhere to $1/2[f(x+0)+f(x-0)]$, and it converges uniformly on closed intervals of continuity. In the generalized Λ-bounded variation case this cannot happen because there exist functions GHBV whose Fourier series diverge in some point of continuity.

It is enough to consider the Lebesgue function defined in $[-\pi,\pi]$ as follows: $f(x) = c_k \sin a_k x$ in $I_k = [\pi/a_k, \pi/a_{k-1}]$ $(k = 1,2,\ldots)$, $f(0) = 0$, $f(-x) = f(x)$, with a_k and c_k suitable constants such that f is continuous and its Fourier series diverges in $x = 0$ ([1], p. 128). This function is GHBV in $[-\pi,\pi]$ as we can see by applying the following

THEOREM 4. Let f be a BV continuous function in $[a,b]$. For each $\Lambda = \{\lambda_i\}$ and $\varepsilon > 0$ there exists a complete subbase $\mathcal{A}$ such that $V_\Lambda(f,\mathcal{A}) \le \varepsilon$.

PROOF. Let $p \in \mathbb{N}$ be such that $V(f) \le \lambda_{p+1}\varepsilon/2$, where $V(f)$ is the variation of f, and let $\delta \in \mathbb{R}_+$ have the property that if $\mu(I) < \delta$ then $|f(I)| \le \varepsilon/\left(2\sum_1^p 1/\lambda_i\right)$.

Let $\mathcal{A}$ be a complete subbase consisting precisely of all intervals $[x,x+\gamma]$, $[x-\gamma,x]$ contained in $[a,b]$, with $0 < \gamma < \delta$ and $x \in [a,b]$. Let $\{I_i\}$ be a sequence of intervals in $\mathcal{A}$; then

$$\sum_1^\infty 1/\lambda_i |f(I_i)| = \sum_1^p 1/\lambda_i |f(I_i)| + \sum_{p+1}^\infty 1/\lambda_i |f(I_i)| \le \varepsilon ;$$

and the proposition follows. ■

BIBLIOGRAPHY

1. N. K. Bary, A Treatise on Trigonometric Series, vol. I, The Macmillan Company, New York, 1964.

2. B. Bongiorno and P. Vetro, "Su un teorema di Riesz", Atti Acc. Scienze Lettere ed Arti di Palermo, serie IV, 37 (1977-78), parte I, 3-13.

3. L. Carleson, "On convergence and growth of partial sums of Fourier series", Acta Math., 116 (1966), 135-157.

4. C. Goffman, "Everywhere convergence of Fourier series", Indiana Univ. Math. J., 20 (1970), 107-113.

5. C. Goffman and D. Waterman, "Functions whose Fourier series converge for every change of variable", Proc. A.M.S. 19 (1968), 80-86.

6. C. Goffman and D. Waterman, "A characterization of the class of functions whose Fourier series converge for every change of variable", J. London Math. Soc. (2) 10 (1975), 69-74.

7. S. Perlman, "Functions of generalized variation", Fund. Math., CV (1980), 199-211.

8. D. Waterman, "On convergence of Fourier series of functions of generalized bounded variation", Studia Math., 44 (1972), 107-117.

9. D. Waterman, "On the summability of Fourier series of functions of Λ-bounded variation", Studia Math., 55 (1976), 87-95.

10. D. Waterman,"On Λ-bounded variation", Studia Math., 57 (1976), 33-45.

11. L. C. Young, "Sur une généralisation de la notion de variation de puissance p-ième bornée au sens de M. Wiener, et sur la convergence des séries de Fourier", Comptes Rendus, 204 (1937), 470-472.

ISTITUTO DI MATEMATICA
VIA ARCHIRAFI 34
90123 PALERMO (ITALY)

Contemporary Mathematics
Volume 42, 1985

ON THE LEVEL SET STRUCTURE OF A CONTINUOUS FUNCTION

B.S. THOMSON

ABSTRACT. By a level of a continuous function f is meant the set $f^{-1}(f(x))$. Two theorems are proved that give an indication of the porosity structure of level sets.

Let f be a continuous real-valued function defined on the real line. By a level or level set of f at a point x we mean the set $f^{-1}(f(x))$. Certainly any such set is closed and given any closed set it may appear as a level for some continuous function. A number of authors, most notably K. Garg, have investigated the level structure of continuous functions (see for example Bruckner [3, Chapter XIII]). In particular we mention the following result of Garg [6]: if f is continuous and nowhere monotone then except for a set of x of the first category each level $f^{-1}(f(x))$ is perfect. In this article we balance Garg's result by proving that these levels must be, for the most part, very "thin".

The thinness concept needed here is the notion of set porosity first introduced by Dolženko [4] and which has been successfully applied to the study of various questions in analysis by a number of authors (eg. [1], [2], [4], and [5]). If $A \subset R$ and $x \in R$ then the right hand porosity of A at x is defined to be the limit superior as $h \to 0+$ of $\lambda(A,x,h)/h$ where $\lambda(A,x,h)$ denotes the length of the largest open interval in $(x,x+h)$ that lies in the complement of A. A set is said to be strongly porous on the right at one of its points x if the right hand porosity of A at x is 1. Left hand versions of these concepts are defined similarly.

Our first theorem states that residually the levels of a continuous function f that is constant in no interval must be strongly porous on both the right and the left at each of their points.

1980 Mathematics Subject Classification. 26A15

THEOREM 1 *Let f be a continuous nowhere constant function. Then except for a set of x of the first category each level $f^{-1}(f(x))$ is strongly porous on both sides at x.*

PROOF. Define X_m to be the set of all points x at which the right hand porosity of $L_x = f^{-1}(f(x))$ is less than $1-1/m$. Then for each $x \in X_m$ there must be a positive number $\delta(x)$ so that whenever $0 < t < \delta(x)$ we must have $\lambda(L_x,x,t) < t(1-1/m)$. If we set $X_{mn} = \{x \in X_m : \delta(x) > 1/n\}$ then the double sequence $\{X_{mn}\}$ includes every point x for which the right hand porosity of L_x at x is not equal to 1. The theorem is proved by showing that each set X_{mn} is nowhere dense; this expresses the set of points x for which L_x has right porosity different from 1 as a first category set. As left hand porosity permits a similar argument the theorem is proved.

Suppose then, in order to obtain a contradiction, that some set X_{mn} is dense in an interval (c,d). We may suppose that $d-c < 1/n$. As f is supposed not to be constant in (c,d) there are points x_1 and y_1 in that interval for which $f(x_1) < f(y_1)$. We may take it that $x_1 < y_1$ as similar arguments would handle the reverse inequality. Define $C = (f(x_1) + f(y_1))/2$ and set x_0 equal to $\sup\{x \in (x_1,y_1) : f(x) < C\}$. Certainly we have $0 < h < 1/n$ where $h = y_1 - x_0$. As X_{mn} is dense in (c,d) and f is continuous there must be a point $z \in X_{mn} \cap (x_0 - \frac{h}{m}, x_0)$ for which $f(z) < C$. Since $z \in X_{mn}$ and $0 < h < 1/n < \delta(z)$ we know that $\lambda(L_z,z,h)$ is less than $(1-\frac{1}{m})h$ so that in particular the interval $(z+\frac{h}{m}, z+h)$ must intersect the level L_z. This means that there must be points x in the interval (x_0,y_1) for which $f(x) = f(z) < C$ and this contradicts our definition of the point x_0. From this contradiction the theorem now follows.

This theorem suggests the following problem: is there an example of a continuous function <u>all</u> of whose level sets are both perfect and strongly porous on both sides at each of their points. Possibly the construction of the Gillis function or Foran's example (cited in [3, p.223]) can be modified to exhibit such a function. We shall instead here give an indication that such functions might exist. In this theorem Φ denotes the Banach space of all continuous 2π-periodic functions furnished with the supremum norm $\|f\| = \sup\{|f(t)| : -\pi \le t \le \pi\}$.

THEOREM 2 *There is a residual subset of the space Φ so that for each function in that set every level is strongly porous on both sides at each of its points.*

PROOF. For rational number p and δ with $\delta > 0$ and $p \in (0,1)$ define the set $F_{p\delta}$ as the collection of all functions $f \in \Phi$ for which there is some point x (depending on f, p and δ) with the property that $\lambda(L_x(f),x,h) \leqq ph$ whenever $0 < h < \delta$. We claim that each such $F_{p\delta}$ is nowhere dense in Φ. Then by forming a union over all such rational numbers p and δ we have exhibited the collection of functions which violate the assertion of the theorem on the right at some point as a first category set in Φ. As the left hand version may be similarly handled the theorem will follow.

Observe firstly that no set $F_{p\delta}$ can contain a neighbourhood in Φ. To see this it is enough to note that any such neighbourhood contains an abundance of functions whose graphs are (nonhorizontal) line segments and no such function can belong to a set $F_{p\delta}$. Consequently we are done if we can show that each set $F_{p\delta}$ is closed in Φ.

Let $\{f_n\}$ be a uniformly convergent sequence of functions from some $F_{p\delta}$, let g be the limit of that sequence, and write $\{x_n\}$ as the sequence of points associated with the corresponding f_n as in the definition of the set $F_{p\delta}$. We may suppose that each x_n lies in the interval $[-\pi,\pi]$ and even, by passing to a subsequence if necessary, that $\{x_n\}$ converges to a point z. We will show that $\lambda(L_z(g),z,h) \leq ph$ for any $0 < h < \delta$ and this will prove that $g \in F_{p\delta}$. Since this holds for any such sequence $F_{p\delta}$ must then be closed and the proof is complete.

Suppose, contradicting this, that $\lambda(L_z(g),z,h) > ph$ for some such number h. Then there is an interval J in $(z,z+h)$ for which $|J| > ph$ and $J \cap L_z(g) = \emptyset$. We can pass then to a closed subinterval $[c,d] \subset J$ with $d-c > ph$ and $|g(x)-g(z)| \geqq \varepsilon > 0$ for every $x \in [c,d]$. For sufficiently large n the interval $[c,d]$ is contained in (x_n,x_n+h), $|f_n(t)-g(t)| < \varepsilon/3$ everywhere, and $|g(z)-f_n(x_n)| < \varepsilon/3$. But then for any x in the interval $[c,d]$ we would have for these n that $|f_n(x)-f_n(x_n)| \geqq \varepsilon/3$. Consequently the level of f_n at x_n for sufficiently large n cannot intersect the interval $[c,d]$; this gives $\lambda(L_{x_n}(f_n),x_n,h) \geqq d-c > ph$ which contradicts the choice of the f_n and the x_n. The theorem now follows.

Acknowledgement. The author wishes to thank Professor Paul Humke for indicating a simplification that has been used in the proof of Theorem 1.

REFERENCES

1. C.L. Belna, M.J. Evans, and P.D. Humke, Most directional cluster sets have common values, Fund. Math. 101 (1978), 1-10.

2. ___________, Symmetric and ordinary differentiation, Proc. Amer. Math. Soc., 72 (1978), 261-267.

3. A.M. Bruckner, Differentiation of real functions, Lecture Notes in Math. #659, Springer (1978).

4. E.P. Dolženko, Boundary properties of arbitrary functions, Math. USSR - Izv. 1 (1967), 1-12.

5. M.J. Evans and P.D. Humke, The equality of unilateral derivates, Proc. Amer. Math. Soc., 79 (1980), 609-613.

6. K.M. Garg, On the level sets of a continuous nowhere monotone function, Fund. Math., 52 (1963), 59-68.

7. J. Gillis, Note on a conjecture of Erdös, Quart. J. Math. Oxford 10 (1939), 151-154.

MATHEMATICS DEPARTMENT
SIMON FRASER UNIVERSITY
BURNABY, B.C.
CANADA V5A 1S6

Contemporary Mathematics
Volume 42, 1985

SOME PROPERTIES OF THE LITTLEWOOD-PALEY g-FUNCTION

Wang Silei

ABSTRACT. For f in $L^\infty(R^n)$ or in $BMO(R^n)$ the magnitude and BMO norm of $g(f)$ are studied.

Let $\mathbb{R}^n$ be the n-dimensional Euclidean space, and let $f(x) \in L^p(\mathbb{R}^n)$, $1 \le p \le \infty$. The g-function of f is defined as follows. Suppose that the Poisson integral of f is

$$u(x,y) = \int_{\mathbb{R}^n} P_y(t) f(x-t)dt = (f * P_y)(x) \qquad (y>0), \tag{1.1}$$

$$P_y(t) = \frac{c_n y}{(|t|^2 + y^2)^{(n+1)/2}}, \quad c_n = \frac{\Gamma\left(\frac{n+1}{2}\right)}{\pi^{(n+1)/2}}. \tag{1.2}$$

We let Δ denote the Laplace operator in $\mathbb{R}^{n+1}_+$, that is $\Delta = \partial^2/\partial y^2 + \sum_{j=1}^{n} \partial^2/\partial x_j^2$; ∇ is the correcponding gradient, i.e.,

$$|\nabla u(x,y)|^2 = \left|\frac{\partial u}{\partial y}\right|^2 + |\nabla_x u(x,y)|^2 = \left|\frac{\partial u}{\partial y}\right|^2 + \sum_{j=1}^{n} \left|\frac{\partial u}{\partial x_j}\right|^2. \tag{1.3}$$

With these notations we define $g(f)(x)$ by

$$g(f)(x) = \left(\int_0^\infty y|\nabla u(x,y)|^2 dy\right)^{1/2}. \tag{1.4}$$

It is easy to see that the g-function of f is well defined, although it may take the value ∞ on some set with positive measure. This definition of $g(f)(x)$, which plays an important role in the Fourier analysis and is a component part of Littlewood-Paley theory, was first introduced by E. M. Stein [3]. Stein proved that, if $p=2$, then $g(f)(x)$ takes finite values almost everywhere and [4, pp. 82-83]

1980 Mathematics Subject Classification。 42B25.

$$\|g(f)(x)\|_2 = \frac{1}{\sqrt{2}} \|f(x)\|_2 ; \tag{1.5}$$

moreover,

$$\|g(f)(x)\|_p \le A_p \|f(x)\|_p , \qquad (1 < p < \infty) \tag{1.6}$$

where the notation $\|f\|_p$ denotes the L_p norm of f, i.e.,

$$\|f\|_p = \left(\int_{\mathbb{R}^n} |f(x)|^p dx\right)^{1/p} . \tag{1.7}$$

The aim of this paper is to study what happens if $p = \infty$. The results we obtain are as follows:

THEOREM 1. There exists function $f_0(x) \in L^\infty(\mathbb{R}^n)$ such that $g(f_0)(x) = \infty$ everywhere.

THEOREM 2. If $f \in L^\infty(\mathbb{R}^n)$ and $g(f)(x) < \infty$ almost everywhere, then $g(f)(x) \in BMO(\mathbb{R}^n)$.

Let f be a locally integrable function on $\mathbb{R}^n$. Then f is said to be of bounded mean oscillation (abbreviated as $BMO(\mathbb{R}^n)$) if

$$\sup \frac{1}{|Q|} \int_Q |f(x) - f_Q| dx = \|f\|_* < \infty , \tag{1.8}$$

where the supremum ranges over all finite cubes Q in $\mathbb{R}^n$ whose sides are parallel to the axes, $|Q|$ is the Lebesgue measure of Q, and f_Q denotes the mean value of f over Q, namely

$$f_Q = \frac{1}{|Q|} \int_Q f(x) dx . \tag{1.9}$$

Note that every bounded function is in $BMO(\mathbb{R}^n)$, i.e., $L^\infty(\mathbb{R}^n) \subset BMO(\mathbb{R}^n)$; however the function $\log|x|$ can be seen to be in $BMO(\mathbb{R}^n)$, so the converse does not hold. This means that $L^\infty(\mathbb{R}^n)$ is a proper subset of $BMO(\mathbb{R}^n)$. Thus, it is natural to ask if we can improve the result in Theorem 2 so as to yield the statement that $g(f)(x) \in L^\infty(\mathbb{R}^n)$.

THEOREM 3. There exists a function $f_0(x) \in L^\infty(\mathbb{R}^n)$ with $g(f_0)(x) < \infty$ almost everywhere, while $g(f_0) \notin L^\infty(\mathbb{R}^n)$. The above Theorem 3 shows that, in a sense, the result in Theorem 2 is best possible. However, Theorem 2' below shows that the conditions imposed on f can be essentially relaxed.

THEOREM 2'. If $f \in BMO(\mathbb{R}^n)$ and $g(f)(x) < \infty$ a.e., then $g(f)(x) \in BMO(\mathbb{R}^n)$; moreover, there exists a constant, depending only on dimension, so that

$$\|g(f)\|_* \le C\|f\|_* . \tag{1.10}$$

As for the values which $g(f)(x)$ takes on, we have the following theorem.

THEOREM 4. Suppose that $f \in BMO(\mathbb{R}^n)$. Then either $g(f)(x) = \infty$ a.e., or we have $g(f)(x) < \infty$ a.e.

2. PROOF OF THEOREM 1. Write

$$f_0(x) = \begin{cases} 1 & (x \in \mathbb{R}^n_+ = \{x = (x_1, x_2, \ldots, x_n) \mid x_n \geq 0\}), \\ 0 & (x \notin \mathbb{R}^n_+). \end{cases} \tag{2.1}$$

Obviously, $f_0 \in L^\infty(\mathbb{R}^n)$. We have to prove that $g(f_0)(x) = \infty$ at each point $x \in \mathbb{R}^n$. By (1.2) and (1.3), the Poisson integral of $f_0(x)$ is

$$u(x,y) = (f_0 * P_y)(x) = \int_{\mathbb{R}^n} f_0(t) P_y(x-t)\,dt,$$

and

$$\frac{\partial u}{\partial x_n}(x,y) = \int_{\mathbb{R}^n} f_0(t) \frac{\partial P_y}{\partial x_n}(x-t)\,dt =$$

$$- (n+1)c_n \int_{\mathbb{R}^n} f_0(x-t) \frac{t_n y}{(|t|^2+y^2)^{(n+3)/2}}\,dt. \tag{2.2}$$

We have two different cases:

(1) $n = 1$. In this case,

$$\frac{\partial u}{\partial x}(x,y) = -2c_1 \int_{-\infty}^{x} \frac{ty}{(t^2+y^2)^2}\,dt = \frac{c_1 y}{x^2+y^2} \qquad (y > 0). \tag{2.3}$$

Thus, by (1.4),

$$g(f_0)(x) = \left\{\int_0^\infty y\,|\nabla u(x,y)|^2 dy\right\}^{1/2} \geq c_1 \left\{\int_0^\infty y^3 (x^2+y^2)^{-2} dy\right\}^{1/2} = \infty, \tag{2.4}$$

(2) $n \geq 2$. Writing

$$t^* = (t_1, t_2, \ldots, t_{n-1}) \in \mathbb{R}^{n-1}, \qquad t = (t^*, t_n),$$

we have by (2.2),

$$\frac{\partial u}{\partial x_n}(x,y) = -(n+1)c_n \int_{\mathbb{R}^{n-1}} dt^* \int_{-\infty}^{\infty} f_0(x-t) \frac{t_n y}{(|t^*|^2 + t_n^2 + y^2)^{(n+3)/2}}\,dt_n =$$

$$= -(n+1)c_n \int_{\mathbb{R}^{n-1}} dt^* \int_{-\infty}^{x_n} \frac{t_n y}{(|t^*|^2 + t_n^2 + y^2)^{(n+3)/2}}\,dt_n =$$

$$= c_n \int_{\mathbb{R}^{n-1}} \frac{y}{(|t^*|^2 + x_n^2 + y^2)^{(n+1)/2}}\,dt^* =$$

$$= \frac{c_n y}{x_n^2 + y^2} \int_{\mathbb{R}^{n-1}} \frac{du}{(1+|u|^2)^{(n+1)/2}} \tag{2.5}$$

Therefore,

$$g(f_0)(x) = \left\{\int_0^\infty y|\nabla u(x,y)|^2 dy\right\}^{1/2} \sim \left\{\int_0^\infty y^3(x_n^2+y^2)^{-2}dy\right\}^{1/2} = \infty . \quad (2.6)$$

Collecting (1) and (2), Theorem 1 follows.

3. A LEMMA

LEMMA. Suppose that $f \in \mathrm{BMO}(\mathbb{R}^n)$. If there exist a cube whose side are parallel to the axes (we will consider these cubes only below) and a point $x_0 \in Q^*$, such that

$$g((f(x) - f_{Q^*})\overline{\chi}_{Q^*})(x_0) < \infty , \quad (3.1)$$

where

$$\overline{\chi}_{Q^*}(x) = 1-\chi_{Q^*}(x) = \begin{cases} 0 & (x \in Q^*) , \\ 1 & (x \in {}^cQ^*,\ {}^cQ^* \text{ being the complement of } Q^*), \end{cases} \quad (3.2)$$

then $g(f(x)) < \infty$ a.e., and $g(f)(x) \in \mathrm{BMO}(\mathbb{R}^n)$; moreover, there is a constant C, depending only on the dimension n, such that the following inequality holds true for each cube Q:

$$\int_Q |g(f)(x) - (g(f))_Q| dx \le C|Q| \, \|f\|_* . \quad (3.3)$$

Proof. The argument is divided into three steps.

(1). Suppose that the edgelength of the cube Q^* is $4h$. Define Q_1 to be the cube with edgelength h, having the same center as Q^*. We first prove that $g(f)(x) < \infty$ a.e. on Q_1.

Write

$$f(x) = f_{Q^*} + (f(x)-f_{Q^*})\chi_{Q^*}(x) + (f(x)-f_{Q^*})\overline{\chi}_{Q^*}(x) = \sum_{i=1}^{3} f_i(x) \quad (3.4)$$

and

$$u_i(x,y) = (f_i * P_y)(x) \qquad (i = 1,2,3) . \quad (3.5)$$

Thus

$$u_1(x,y) \equiv f_{Q^*} = f_1(x) ,$$

so that

$$g(f_1)(x) \equiv 0 . \quad (3.6)$$

As for $f_2(x)$, we have by (1.5), that

$$\int_{Q_1} |g(f_2)(x)|^2 dx \le \int_{\mathbb{R}^n} |g(f_2)(x)|^2 dx = \frac{1}{2}\int_{\mathbb{R}^n} |f_2(x)|^2 dx =$$

$$= \frac{1}{2}\int_{Q^*} |f(x)-f_{Q^*}|^2 dx \le A|Q| \cdot \|f\|_*^2 \quad \text{(A being a constant).} \quad (3.7)$$

The last inequality is an easy consequence of the well known property of BMO function established by John and Nirenberg [2]

$$|\{x\in Q \mid |f(x)-f_Q|>\alpha\}| \le e^{-c\alpha/\|f\|_*}|Q| \qquad (\forall\ \alpha>0). \tag{3.8}$$

Thus by (3.7)

$$\int_Q |g(f_2)(x)|dx \le |Q|^{\frac{1}{2}}\Big(\int_Q |g(f_2)(x)|^2 dx\Big)^{\frac{1}{2}} \le A|Q|\cdot\|f\|_* . \tag{3.9}$$

It also follows from (3.7) that

$$g(f_2)(x)<\infty \tag{3.10}$$

a.e. on $\mathbb{R}^n$.

Now we consider $g(f_3)(x)$, $x\in Q_1$. In virtue of

$$g(f_3)(x)=\Big(\int_0^\infty y|\nabla u_3(x,y)|^2 dy\Big)^{1/2} \le$$

$$\le\Big(\int_0^\infty y|\nabla u_3(x,y) - u_3(x_0,y)|^2 dy\Big)^{1/2} + g(f_3)(x_0)$$

and the hypothesis that $g(f_3)(x_0)<\infty$, it is sufficient to prove that

$$I(x)\equiv\Big(\int_0^\infty y|\ u_3(x,y) - u_3(x_0,y)|^2 dy\Big)^{1/2}<\infty, \tag{3.11}$$

By Hölder's inequality,

$$I(x)\le\Big(\int_0^\infty y\Big|\frac{\partial u_3}{\partial y}(x,y)-\frac{\partial u_3}{\partial y}(x_0,y)\Big|^2 dy\Big)^{1/2}+$$

$$+\sum_{j=1}^n\Big(\int_0^\infty y\Big|\frac{\partial u_3}{\partial x_j}(x,y)-\frac{\partial u_3}{\partial x_j}(x_0,y)\Big|^2 dy\Big)^{1/2}=I_0(x)+\sum_{j=1}^n I_j(x). \tag{3.12}$$

Now we estimate $I_0(x)$ as follows:

$$I_0(x)=\Big(\int_0^\infty y\Big|\frac{\partial u_3}{\partial y}(x,y)-\frac{\partial u_3}{\partial y}(x_0,y)\Big|^2 dy\Big)^{1/2}\le I_{0,1}(x)+I_{0,2}(x), \tag{3.13}$$

where

$$I_{0,1}(x)=\Big(\int_0^h y\Big|\frac{\partial u_3}{\partial y}(x,y)-\frac{\partial u_3}{\partial y}(x_0,y)\Big|^2 dy\Big)^{1/2} \tag{3.14}$$

and

$$I_{0,2}(x)=\Big(\int_h^\infty y\Big|\frac{\partial u_3}{\partial y}(x,y)-\frac{\partial u_3}{\partial y}(x_0,y)\Big|^2 dy\Big)^{1/2}. \tag{3.15}$$

It is easy to verify that

$$\frac{\partial u_3}{\partial y}(x,y)-\frac{\partial u_3}{\partial y}(x_0,y)=$$

$$=c_n\int_{\mathbb{R}^n}\Big(\frac{|x-t|^2-ny^2}{(|x-t|^2+y^2)^{(n+3)/2}}-\frac{|x_0-t|^2-ny^2}{(|x_0-t|^2+y^2)^{(n+3)/2}}\Big)f_3(t)dt. \tag{3.16}$$

Hence

$$I_{0,1}(x) = c_n\Big\{\int_0^h y\Big|\int_{^cQ^*}\Big(\frac{|x-t|^2 - ny^2}{(|x-t|^2+y^2)^{(n+3)/2}} -$$

$$- \frac{|x_0-t|^2 - ny^2}{(|x_0-t|^2+y^2)^{(n+3)/2}}\Big)(f(t)-f_{Q^*})dt\Big|^2 dy\Big\}^{1/2}$$

$$\leq c_n\Big\{\int_0^h y\Big[\int_{^cQ^*}\Big(\Big|\frac{|x-t|^2 - ny^2}{(|x-t|^2+y^2)^{(n+3)/2}}\Big| +$$

$$+ \Big|\frac{|x_0-t|^2 - ny^2}{(|x_0-t|^2+y^2)^{(n+3)/2}}\Big|\Big)|f(t)-f_{Q^*}|dt\Big]^2 dy\Big\}^{1/2}$$

$$\leq A_n\Big\{\int_0^h y\Big[\int_{^cQ^*}\Big(\frac{1}{(|x-t|+y)^{(n+1}} +$$

$$+ \frac{1}{(|x_0-t|+y)^{n+1}}\Big)|f(t)-f_{Q^*}|dt\Big]^2 dy\Big\}^{1/2}. \tag{3.17}$$

A_n being a constant, not necessarily the same at each occurrence, depending only on the dimension n. Noticing that the condition $x \in Q_1$ and $t \in {}^cQ^*$ implies that

$$|x-t| \geq A_n|x_0-t| \geq A_n(|x_0-t|+h) \tag{3.18}$$

and

$$\frac{1}{(|x-t|+y)^{n+1}} \leq \frac{A_n}{|x_0-t|^{n+1} + h^{n+1}}, \tag{3.19}$$

from (3.17) and (3.19) we obtain that

$$I_{0,1}(x) \leq A_n\Big\{\int_0^h y\Big(\int_{^cQ^*}\frac{|f(t)-f_{Q^*}|}{|x_0-t|^{n+1} + h^{n+1}}dt\Big)^2 dy\Big\}^{1/2}$$

$$\leq A_n\Big(\int_0^h y\cdot h^{-2}\|f\|_*^2\,dy\Big)^{1/2} \leq A_n\|f\|_* \tag{3.20}$$

since a BMO function must satisfy the well known inequality of Fefferman and Stein

$$\int_{\mathbb{R}^n}\frac{|f(t)-f_{Q_d}|}{d^{n+1} + |t-t_0|^{n+1}}dt \leq \frac{A_n}{d}\|f\|_* \qquad (d > 0),$$

Q_d being a cube with edge length d and center t_0 [1, p. 142]. We observe that the same method used in [1] gives the following generalized inequality

$$\int_{\mathbb{R}^n} \frac{|f(t)-f_{Q_d}|}{d^{n+\alpha}+|t-t_0|^{n+\alpha}}\,dt \le \frac{A_n}{d^{\alpha}}\|f\|_* \qquad (\alpha>0)\,. \tag{3.21}$$

Now let us estimate $I_{0,2}(x)$. First we note that, if $x \in Q_1$, then

$$\left|\frac{|x-t|^2-ny^2}{(|x-t|^2+y^2)^{(n+3)/2}}-\frac{|x_0-t|^2-ny^2}{(|x_0-t|^2+y^2)^{(n+3)/2}}\right|$$

$$\le A_n|x-x_0|\,\frac{|\tilde{x}-t|((n+2)y^2+|\tilde{x}-t|^2)}{(|\tilde{x}-t|^2+y^2)^{(n+5)/2}} \le A_n\,\frac{h}{|x_0-t|^{n+2}+y^{n+2}}\,, \tag{3.22}$$

since $t \in {}^cQ^*$ and $\tilde{x} \in Q_1$, $\tilde{x}$ being a point lying on the segment $x\,x_0$. Considering the conditions (3.15) and (3.16), (3.21), we get

$$I_{0,2}(x) \le A_n\left(\int_h^{\infty} y\left[\int_{^cQ^*}\frac{h|f(t)-f_{Q^*}|}{|x_0-t|^{n+2}+y^{n+2}}\,dt\right]^2 dy\right)^{1/2}$$

$$\le A_n\left(\int_h^{\infty} y\left[\int_{^cQ^*}\frac{h|f(t)-f_{Q^*}|}{\left(|x_0-t|^{n+\frac{1}{2}}+h^{n+\frac{1}{2}}\right)y^{3/2}}\,dt\right]^2 dy\right)^{1/2}$$

$$\le A_n\left(\int_h^{\infty} hy^{-2}\|f\|_*^2 dy\right)^{1/2} \le A_n\|f\|_*\,. \tag{3.23}$$

Combining (3.13), (3.20) and (3.23), we get the estimate

$$I_0(x) \le A_n\|f\|_*\,. \tag{3.24}$$

The estimates for $I_j(x)$ $(j=1,2,\ldots,n)$ are similar. In fact, by (3.12) we have

$$I_j(x) = \left(\int_0^{\infty} y\left|\frac{\partial u_3}{\partial x_j}(x,y)-\frac{\partial u_3}{\partial x_j}(x_0,y)\right|^2 dy\right)^{1/2} \le I_{j,1}(x)+I_{j,2}(x), \tag{3.25}$$

$$I_{j,1}(x) = \left(\int_0^{h} y\left|\frac{\partial u_3}{\partial x_j}(x,y)-\frac{\partial u_3}{\partial x_j}(x_0,y)\right|^2 dy\right)^{1/2}, \tag{3.26}$$

$$I_{j,2}(x) = \left(\int_h^{\infty} y\left|\frac{\partial u_3}{\partial x_j}(x,y)-\frac{\partial u_3}{\partial x_j}(x_0,y)\right|^2 dy\right)^{1/2}. \tag{3.27}$$

It is easy to verify that

$$\frac{\partial u_3}{\partial x_j}(x,y)-\frac{\partial u_3}{\partial x_j}(x_0,y) = -c_n\int_{\mathbb{R}^n}\left(\frac{(n+1)(x_j-t_j)y}{(|x-t|^2+y^2)^{(n+3)/2}} - \right.$$

$$\left. - \frac{(n+1)(x_{0,j}-t_j)y}{(|x_0-t|^2+y^2)^{(n+3)/2}}\right) f_3(t)dt\,. \tag{3.28}$$

Therefore,

$$I_{j,1}(x) = (n+1)c_n\Big(\int_0^h y\Big|\int_{{}^cQ^*}\Big[\frac{(x_j-t_j)y}{(|x-t|^2+y^2)^{(n+1)/2}} - \frac{(x_{0,j}-t_j)y}{(|x_0-t|^2+y^2)^{(n+3)/2}}\Big][f(t)-f_{Q^*}]dt\Big|^2 dy\Big)^{1/2}$$

$$\le (n+1)c_n\Big(\int_0^h y\Big|\int_{{}^cQ^*}\Big[\frac{1}{(|x-t|^2+y^2)^{(n+1)/2}} + \frac{1}{(|x_0-t|^2+y^2)^{(n+1)/2}}\Big][f(t)-f_{Q^*}]dt\Big|^2 dy\Big)^{1/2}$$

$$\le A_n\Big\{\int_0^h y\Big(\int_{{}^cQ^*}\frac{|f(t)-f_{Q^*}|}{h^{n+1}+|x_0-t|^{n+1}}dt\Big)^2 dy\Big\}^{1/2}$$

$$\le A_n\Big\{\int_0^h y\cdot h^{-2}\|f\|_*^2 dy\Big\}^{1/2} \le A_n\|f\|_* \quad (j=1,2,\ldots,n), \tag{3.29}$$

where we also have used the inequalities (3.18) and (3.21).

The estimates for $I_{j,2}(x)$ $(j=1,2,\ldots,n)$ are similar to that of $I_{0,2}(x)$, if we note that

$$\frac{\partial^2 P_y}{\partial x_j^2}(x) = \frac{c_n(n+1)y[(n+3)x_j^2 - (|x|^2+y^2)]}{(|x|^2+y^2)^{(n+5)/2}} \quad (j=1,2,\ldots,n) \tag{3.30}$$

and the conditions $x\in Q_1$, $t\in {}^cQ^*$ imply that

$$(n+1)c_n\Big|\frac{(x_j-t_j)y}{(|x-t|^2+y^2)^{(n+3)/2}} - \frac{(x_{0,j}-t_j)y}{(|x_0-t|^3+y^2)^{(n+3)/2}}\Big|$$

$$\le A_n|x-x_0|\,\frac{y[(n+3)(\tilde{x}_j-t_j)^2+|\tilde{x}-t|^2+y^2]}{(|\tilde{x}-t|^2+y^2)^{(n+5)/2}} \le \frac{A_n h}{|x_0-t|^{n+2}+y^{n+2}}, \tag{3.31}$$

$\tilde{x}$ being a point lying on the segment $\overline{x\,x_0}$. Hence

$$I_{j,2}(x) = \Big(\int_h^\infty y\Big|\frac{\partial u_3}{\partial x_j}(x,y) - \frac{\partial u_3}{\partial x_j}(x_0,y)\Big|^2 dy\Big)^{1/2}$$

$$\le A_n\Big(\int_h^\infty y\Big[\int_{{}^cQ^*}\frac{h|f(t)-f_{Q^*}|}{|x_0-t|^{n+2}+y^{n+2}}dt\Big]^2 dy\Big)^{1/2}$$

$$\le A_n\Big(\int_h^\infty y\Big[\int_{{}^cQ^*}\frac{h|f(t)-f_{Q^*}|}{\big(|x_0-t|^{n+\frac12}+h^{n+\frac12}\big)y^{3/2}}dt\Big]^2 dy\Big)^{1/2}$$

$$\leq A_n\left(\int_h^\infty h\cdot y^{-2}\|f\|_*^2\,dy\right)^{1/2} \leq A_n\|f\|_* \qquad (j=1,2,\ldots,n). \tag{3.32}$$

Combining (3.25), (3.29) and (3.32), we get

$$I_j(x) \leq A_n\|f\|_* \qquad (j=1,2,\ldots,n). \tag{3.33}$$

Substituting (3.33), (3.24) into (3.12), we obtain

$$I(x) \leq A_n\|f\|_* , \tag{3.34}$$

so that

$$g(f_3)(x) \leq A_n\|f\|_* + g(f_3)(x_0) . \tag{3.35}$$

Therefore

$$g(f)(x) \leq g(f_2)(x) + g(f_3)(x) < \infty \qquad \text{a.e. on } Q_1 ,$$

since $g(f_2)(x) < \infty$ a.e. on $\mathbb{R}^n$. This completes the proof of (1).

(2). We will prove that $g(f)(x)$ is finite a.e. on $\mathbb{R}^n$. It suffices to prove that, $g(f)(x) < \infty$ a.e. on each cube. Without loss of generality, we may assume that $Q \supset Q^*$. Similarly to (1), we have

$$f(x) = f_{Q_2} + (f(x)-f_{Q_2})\cdot\chi_{Q_2}(x) + (f(x)-f_{Q_2})\overline{\chi}_{Q_2}(x) = \sum_1^3 f_i(x) , \tag{3.36}$$

with the cube Q_2 concentric with Q but with edge expanded four times. $g(f_1)(x) \equiv 0$, since f_1 is a constant; $g(f_2)(x) < \infty$ a.e., since $f_2 \in L^2(\mathbb{R}^n)$. By (1), $g(f)(x)$ is finite on a subset $Q^* \subset Q_2$ with positive measure. Therefore, by the well known inequality

$$g(f_3)(x) \leq g(f_2)(x) + g(f)(x) ,$$

it follows that, there exists a point $\overline{x} \in Q_2$ so that $g((f(x)-f_{Q_2})\cdot\overline{\chi}_{Q_2})(\overline{x}) < \infty$. Again by (1), $g(f)(x)$ is finite a.e. on Q. This completes the proof of (2).

(3). At this stage we will prove that there exists a constant C, depending only on the dimension n, so that (3.3) holds.

Let Q be any cube with edge length h. Without loss of generality we may assume that its center is 0. Let Q^* denote the cube with the same center as Q but with edge length $4h$. As in (1), we have

$$f(x) = f_{Q^*} + (f(x)-f_{Q^*})\cdot\chi_{Q^*}(x) + (f(x)-f_{Q^*})\cdot\overline{\chi}_{Q^*}(x) = \sum_1^3 f_i(x) \tag{3.36'}$$

and

$$g(f_1)(x) \equiv 0 . \tag{3.37}$$

By (3.9),

$$\int_Q |g(f_2)(x)|dx \le A_n|Q|\cdot\|f\|_* . \tag{3.38}$$

On the other hand, it has been shown in (2) that $g(f)(x) < \infty$ a.e., so that $g(f_3)(x) < \infty$ a.e. Therefore, it must exist an point $\tilde{x} \in Q$ for which $g(f_3)(\tilde{x}) < \infty$. In proving (1) it has been shown (see (3.11) and (3.34)) that

$$I(x) = \left(\int_0^\infty y|\nabla u_3(x,y)-\nabla u_3(\tilde{x},y)|^2 dy\right)^{1/2} \le A_n\|f\|_*$$

for $x \in Q$. Thus for $x \in Q$,

$$|g(f_3)(x)-g(f_3)(\tilde{x})| \le \left(\int_0^\infty y|\nabla u_3(x,y)-\nabla u_3(\tilde{x},y)|^2 dy\right)^{1/2} \le A_n\|f\|_*$$

so

$$\int_Q |g(f_3)(x)-g(f_3)(\tilde{x})|dx \le A_n|Q|\,\|f\|_* . \tag{3.39}$$

Therefore by (3.38) and (3.39),

$$\int_Q |g(f)(x)-g(f_3)(\tilde{x})|dx = \int_Q |g(f_2+f_3)(x)-g(f_3)(x)+g(f_3)(x)-g(f_3)(\tilde{x})|dx$$

$$\le \int_Q |g(f_2)(x)|dx + \int_Q |g(f_3)(x)-g(f_3)(\tilde{x})|dx \le A_n|Q|\cdot\|f\|_* . \tag{3.40}$$

Again by (3.39), we have finally

$$\int_Q |g(f)(x)-(g(f))_Q|dx \le \int_Q |g(f)(x)-g(f_3)(\tilde{x})|dx + \int_Q |g(f_3)(\tilde{x})-g(f))_Q|dx$$

$$\le A_n|Q|\cdot\|f\|_* + |Q|\cdot|g(f_3)(\tilde{x})-(g(f))_Q|$$

$$\le A_n|Q|\cdot\|f\|_* + \int_Q |g(f_3)(x)-g(f_3)(\tilde{x})|dx$$

$$\le A_n|Q|\cdot\|f\|_* + A_n|Q|\cdot\|f\|_* \le A_n|Q|\cdot\|f\|_* . \tag{3.41}$$

This completes the proof of (3), and, therefore, that of the lemma.

4. PROOF OF THEOREMS 2, 2' AND 4.

Since Theorem 2' generalizes Theorem 2, it suffices to prove Theorem 2'. Let Q be any cube, then we have the decomposition

$$f(x) = f_{Q^*} + (f(x)-f_{Q^*})\cdot\chi_{Q^*}(x) + (f(x)-f_{Q^*})\cdot\overline{\chi}_{Q^*}(x) = \sum_1^3 f_i(x) , \tag{4.1}$$

Q^* being a cube with the same center as Q but expanded four times. Under the hypothesis,

$$g(f_3)(x) < \infty \quad \text{a.e. on } \mathbb{R}^n .$$

Thus Theorem 2' follows from the Lemma.

The proof of Theorem 4 is similar to that of Theorem 2'. We omit the details.

5. PROOF OF THEOREM 3.

There are two different cases; (1) $n=1$; (2) $n\geq 2$.

(1). The case $n=1$. Putting

$$f_0(x) = \begin{cases} 1 & x\in(0,1)\,, \\ 0 & x\notin(0,1)\,, \end{cases} \tag{5.1}$$

then obviously $f_0\in L^2(\mathbb{R})$. By a well known result introduced in §1, we have $g(f_0)(x)<\infty$ a.e. On the other hand,

$$\frac{\partial u}{\partial x}(x,y) = -2c_1\int_{-\infty}^{\infty} f_0(x-t)\frac{ty}{(t^2+y^2)^2}\,dt = -2c_1 y\int_{x-1}^{x}\frac{t}{(t^2+y^2)^2}\,dt =$$

$$= c_1 y\,\frac{1-2x}{(x^2+y^2)[(x-1)^2+y^2]}\,.$$

Thus if $x\to 0-$,

$$(g(f_0)(x))^2 \geq \int_0^1 y\left|\frac{\partial u}{\partial x}(x,y)\right|^2 dy \geq A\int_0^1 y^3(x^2+y^2)^{-2}\,dy =$$

$$= A\int_0^{1/|x|} u^3(1+u^2)^{-2}\,du \to \infty\,.$$

This shows that $g(f_0)(x)$ is unbounded in the neighborhood of $x=0$.

(2). $n\geq 2$. Putting

$$f_0(x_1,x_2,\dots,x_n) = f_0(x^*,x_n) = \begin{cases} 1 & x\in T=\{(x^*,x_n): |x^*|<1,\ 0<x_n<1\}\,, \\ 0 & x\notin T\,, \end{cases}$$

then obviously $f_0\in L^2(\mathbb{R}^n)$. By the same reason as in (1), $g(f_0)<\infty$ a.e. On the other hand,

$$\frac{\partial u}{\partial x_n}(x,y) = -(n+1)c_n\int_{\mathbb{R}^{n-1}} f_0(x-t)\frac{t_n y}{(|t|^2+y^2)^{(n+3)/2}}\,dt =$$

$$= -(n+1)c_n\int_{|x^*-t^*|<1} dt^*\int_{x_n-1}^{x_n}\frac{t_n y}{(|t^*|^2+t_n^2+y^2)^{(n+3)/2}}\,dt_n =$$

$$= c_n y\int_{|x^*-t^*|<1}\left\{\frac{1}{(|t^*|^2+x_n^2+y^2)^{(n+1)/2}} - \right.$$

$$\left. - \frac{1}{(|t^*|^2+(x_n-1)^2+y^2)^{(n+1)/2}}\right\} dt^*\,,$$

so that if $x_n<0$,

$$\frac{\partial u}{\partial x_n}((0,x_n),y) = A_n y\left\{\frac{1}{x_n^2+y^2}\int_0^{1/(x_n^2+y^2)^{\frac{1}{2}}} \frac{\gamma^{n-2}}{(1+\gamma^2)^{(n+1)/2}}d\gamma - \right.$$

$$\left. - \frac{1}{(x_n-1)^2+y^2}\int_0^{1/[(x_n-1)^2+y^2]^{\frac{1}{2}}} \frac{\gamma^{n-2}}{(1+\gamma^2)^{(n+1)/2}}d\gamma\right\} \geq$$

$$\geq A_n y \frac{1-2x_n}{(x_n^2+y^2)(y^2+(x_n-1)^2)}\int_0^{1/\sqrt{x_n^2+y^2}} \frac{\gamma^{n-2}}{(1+\gamma^2)^{(n+1)/2}}\, d\gamma\,.$$

Hence if $x_n \to 0-$, we have

$$g(f_0)((0,x_n)) \geq A_n \int_0^{1/|x_n|} u^3(1+u^2)^{-2}\,du \to \infty,$$

which also shows that $g(f_0)(x)$ is unbounded in the neighborhood of $x=0$.

This completes the proof of Theorem 4.

BIBLIOGRAPHY

1. Fefferman, C. L. and E. M. Stein, "H^p spaces of several variables," Acta Math., 129 (1972), 137-193.

2. John, F. and L. Nirenberg, "On functions of bounded mean oscillation," Comm. Pure Appl. Math., 14 (1961), 415-426.

3. Stein, E. M., "On the functions of Littlewood-Paley, Lusin and Marcinkiewicz," Trans. Amer. Math. Soc., 88 (1958), 430-466.

4. Stein, E. M., Singular Integrals and Differentiability Properties of Functions, Princeton University Press, Princeton, 1970.

DEPARTMENT OF MATHEMATICS
HANGZHOU UNIVERSITY
HANGZHOU, PEOPLES REPUBLIC OF CHINA

Contemporary Mathematics
Volume 42, 1985

CHANGE-OF-VARIABLE INVARIANT CLASSES OF FUNCTIONS AND CONVERGENCE OF FOURIER SERIES

Daniel Waterman

ABSTRACT. The condition which characterizes the class of continuous functions whose Fourier series converge everywhere for every change of variable is shown to suffice for regulated functions. The equivalence of various formulations of this condition is extended to regulated functions. A uniform version of this condition is shown to characterize the functions for which uniform convergence of the Fourier series is preserved. This class is also characterized by means of a variational property. The inclusion relations between these classes and various classes of functions of generalized bounded variation are studied.

Let us consider real functions of period 2π. By a change of variable in such a function f we mean its composition $f\circ g$ with a homeomorphism of $[-\pi,\pi]$ with itself. The class of functions which are continuous and whose Fourier series converge everywhere for every change of variable has been characterized by Goffman and Waterman [5]. Goffman remarked in [4] that this characterization was valid for regulated functions, i.e., functions with right and left limits at each point, and that fact was used later by Goffman and Waterman [6] in the characterization of all real functions whose Fourier series converge everywhere for all changes of variable. The sufficiency of the condition of [5] in the case of regulated functions does not follow from the argument given there. In §1 we establish the sufficiency of the Goffman-Waterman condition (GW) for regulated functions. We also show that the equivalence of certain other conditions to GW shown in [2,5] for continuous functions can be extended to regulated functions.

Baernstein and Waterman [2] gave a condition (UGW) which characterized the functions (necessarily continuous) for which uniform convergence is preserved under all changes of variable.

1980 Mathematics Subject Classification. 26A45, 42A20.

In §1 we also show that the UGW condition is a "uniform GW condition."

In §2 we consider the classes of functions satisfying the GW and UGW conditions and various classes of functions of generalized bounded variation which are also change-of-variable invariant. The inclusion relations between these classes are studied in detail and an example is given of a continuous function which satisfies the GW condition but is not of ordered harmonic bounded variation. We also show that UGW can be characterized by a variational property.

1. The condition of Goffman and Waterman (GW) is stated in terms of <u>systems of intervals</u> at a point. For each n, let I_{ni}, $i=1,\ldots,k_n$, be disjoint closed intervals indexed from left to right, i.e., I_n, $I_{n,i-1}$ is to the left of I_{ni}. Let there be a real x such that for every $\delta > 0$ there is an N such that $I_{ni} \subset (x, x+\delta)$ whenever $n > N$. Then the collection $\{I_{ni}\}$, $n=1,2,\ldots$, $i=1,\ldots,k_n$ is called a <u>right system</u> of intervals at x. A left system is defined similarly, $I_{ni} \subset (x-\delta, x)$ and indexed from right to left.

If I denotes an interval $[a,b]$, we write $f(I) = f(b)-f(a)$.

The <u>condition</u> GW <u>is</u>

$$\lim_{n\to\infty} \sum_{i=1}^{k_n} i^{-1} f(I_{ni}) = 0$$

<u>for every right and left system</u>.

In [5] it was observed that for continuous functions the requirement in GW that the I_{ni} be <u>disjoint</u> can be relaxed to that they be <u>nonoverlapping</u>. We will now show that this is true for <u>regulated</u> functions as well.

Suppose $I_i = [a_i, b_i]$ is indexed from left to right and $\Sigma\, i^{-1} f(I_i) > \Delta > 0$. We will now show that if some I_i have endpoints in common, we can form a disjoint collection, indexed in the same direction, for which the corresponding sum exceeds $\Delta/2$.

Suppose $b_i = a_{i+1}$ for some i. At such points shrink the intervals away from the common point by $\varepsilon > 0$ to form intervals I_i'. If I_i has no points in common with adjacent intervals, let $I_i' = I_i$. Noting that if $I_i' \neq I_i$ then the values corresponding to the altered endpoints may be replaced by the limits from the interior plus $o(1)$ as $\varepsilon \to 0$, we see that

$$\Sigma\, i^{-1} f(I_i) = \Sigma\, i^{-1} f(I_i') + \Sigma^* \frac{(i+1/2)(f(b_i+)-f(b_i-))}{i(i+1)} + o(1)$$

as $\epsilon \to 0$, * denoting summation over those i such that $b_i = a_{i+1}$. For regulated f we shall assume that $f(x) = \frac{1}{2}[f(x+)+f(x-)]$ for every x.

For small enough ϵ, at least one of the sums on the right exceeds $\Delta/2$. If it is the first, then $\{I_i'\}$ is our collection of disjoint intervals. Otherwise, if i_j, $j = 1,2,\ldots,$ are the indices appearing in the second sum,

$$1/j \geq 1/i_j > (i_j + \tfrac{1}{2})/i_j(i_j + 1) .$$

If we choose η sufficiently small and set $J_j = [b_{i_j} - \eta, b_{i_j} + \eta\}$, $\{J_j\}$ will be a disjoint collection of intervals indexed in the same direction as $\{I_i\}$ and

$$\Sigma\, j^{-1} f(J_j) > \Delta/2 .$$

In [2] we gave, without proof, another form of the GW condition for continuous functions. Here we will show that a seemingly slightly weaker condition is equivalent to GW for regulated functions. We note that it is easy to see that condition GW implies regulated.

THEOREM 1. A regulated f satisfies condition GW if and only if for every $\epsilon > 0$ and x there is a $\delta > 0$ such that for every finite sequence $\{I_i\}$ of nonoverlapping intervals indexed from left to right (right to left) in $(x,x+\delta)$ $((x-\delta,x))$ with $\cup I_i \subset (x,x+\delta)$ $(\cup I_i \subset (x-\delta,x))$ we have

$$|\Sigma\, i^{-1} f(I_i)| < \epsilon .$$

PROOF. If the GW condition fails at x for a right system, then $\limsup_{n\to\infty} |\Sigma_i\, i^{-1} f(I_{ni})| = 2\epsilon > 0$. Since $\sup \cup_i I_{ni} \to x$ as $n \to \infty$, the condition of the theorem fails.

On the other hand, if there is an x, an $\epsilon > 0$, a sequence $\delta_n \downarrow 0$ and a sequence of sets of nonoverlapping intervals I_{ni}, $i = 1,\ldots,k_n$, indexed from left to right (right to left) such that $\cup_i I_{ni} \subset (x,x+\delta_n)$ $(\subset (x-\delta_n,x))$ and $|\Sigma_i\, i^{-1} f(I_{ni})| \geq \epsilon$, then $\{I_{ni}\}$ is a system for which the GW condition fails, in view of the preceding remarks.■

We will now show that the result on everywhere convergence can be extended to regulated functions.

THEOREM 2. Let f be a regulated function of period 2π. Then $f \circ g$ has an everywhere convergent Fourier series for every homeomorphism g of $[-\pi,\pi]$ with itself if and only if f satisfies the GW condition.

PROOF. The necessity of the GW condition was shown in [5] for continuous functions. Thus we need only show sufficiency.

For any $\delta>0$, if S_n is the n-th partial sum of the Fourier series of f, then

$$(1)\quad S_n(x)-f(x) = \frac{1}{\pi}\int_0^\delta (f(x+t)-f(x+))\frac{\sin nt}{t}\,dt + \frac{1}{\pi}\int_0^\delta (f(x-t)-f(x-))\frac{\sin nt}{t}\,dt + o(1)$$

as $n\to\infty$ uniformly in x. Consider the first integral.

We have shown [10, pp.122-123] that for given $\epsilon>0$ there is an $n(\epsilon,\delta)$ such that $n>n(\epsilon,\delta)$ implies

$$(2)\quad \left|\int_0^\delta (f(x+t)-f(x+))\frac{\sin dt}{t}\right| < \epsilon + \int_0^\delta \left|\sum_1^{N}{}^{*} k^{-1}[f(x+(t+k\pi)/n)-f(x+(t+(k+1)\pi)/n)]\right|dt.$$

Here * denotes summation over odd indices and N is the greatest odd number less than $[n\delta/\pi]-1$.

A similar inequality holds for the other integral in (1).

If f satisfies the equivalent form of GW given in Theorem 1, the integrand on the right in (2) can be made small <u>uniformly</u> in t and n by choosing δ small. Choose δ so that this integral and the one obtained from the other integral in (2) are each less than ϵ in absolute value. Then there is an $n(\epsilon)$ such that

$$|S_n(x)-f(x)| < 5\epsilon/\pi$$

if $n>n(\epsilon)$. ■

The condition UGW of Baernstein and Waterman which is necessary and sufficient for preservation of uniform convergence is that f <u>be continuous</u>, <u>satisfy</u> GW, <u>and</u>

$$\lim_{n\to\infty}\sum_{i=1}^{k_n}(k_n+1-i)^{-1}f(I_{ni}) = 0$$

<u>for every system</u>.

We show next that UGW is equivalent to "GW uniformly in x".

THEOREM 3. f satisfies the UGW condition if and only if for every $\epsilon>0$ there is a $\delta>0$ such that for every x and every finite sequence $\{I_i\}$ of nonoverlapping intervals indexed from left to right (right to left) with $\cup I_i\subset(x,x+\delta)$ $(\cup I_i\subset(x-\delta,x))$ we have

$$|\Sigma\, i^{-1} f(I_i)| < \epsilon .$$

PROOF. In [2] we showed that UGW is equivalent to:

Given $\epsilon > 0$, there exists a $\delta > 0$ such that $\Sigma\, i^{-1}|f(I_i)| < \epsilon$ whenever $\{I_i\}$ is a finite sequence of nonoverlapping intervals indexed in either direction and satisfying $\operatorname{diam}(\cup I_i) < \delta$.

If f satisfies this condition, then given $\epsilon > 0$, using the δ this condition provides, f satisfies the condition of the theorem at every x.

Suppose now that f does not satisfy the UGW condition above. Then there is an $\epsilon > 0$ such that for any $\delta > 0$ there is a finite sequence $\{I_i\}$ of nonoverlapping intervals indexed in one direction or the other, say left to right, so that $\Sigma\, i^{-1}|f(I_i)| < 2\epsilon$ and $\operatorname{diam}(\cup I_i) < \delta$. Fix δ and let $a = \inf \cup I_i$, $b = \sup \cup I_i$; then $b-a < \delta$. Choose $x \in (b-\delta, a)$. Then $\cup I_i \subset (x, x+\delta)$. If $f^+(I) = \max(0, f(I))$ and $f^-(I) = \max(0, -f(I))$, then at least one of $\Sigma\, i^{-1} f^+(I_i)$ and $\Sigma\, i^{-1} f^-(I_i)$ exceeds ϵ, say the first. Consider the I_i for which $f(I_i) > 0$ and index them from left to right to obtain $\{J_i\}$. Then $\Sigma\, i^{-1} f(J_i) > \epsilon$ and $\cup J_i \subset (x, x+\delta)$. Since this may be done for any $\delta > 0$, f does not satisfy the condition of the theorem.■

2. The class of functions of bounded variation (BV) is invariant under change of variable. If X denotes a function class, let X_C denote the continuous functions in X. Let GW and UGW now denote the classes of functions satisfying the GW and UGW conditions respectively. The Dirichlet-Jordan theorem tells us that $BV \subset GW$ and $BV_C \subset UGW$. This theorem has been generalized by replacing BV with larger change-of-variable- invariant classes.

Salem [7] showed that the ΦBV classes of L. C. Young for which the complementary function Ψ satisfies $\Sigma\, \Psi(n^{-1}) < \infty$ may be used. Baernstein [1] showed that there is a function in ΦBV whose Fourier series diverges if $\Sigma\, \Psi(n^{-1}) = \infty$; thus $\Phi BV_C \subset UGW$ if and only if $\Sigma\, \Psi(n^{-1}) < \infty$.

Waterman has shown that the Dirichlet-Jordan theorem can be extended to the class of functions of harmonic bounded variation (HBV) and $HBV \supset \Phi BV$ if Salem's condition is satisfied [8,10].

Let $\{I_n\}$ be a sequence of nonoverlapping intervals in $[-\pi, \pi]$.

We say that $f \in HBV$ if there is an $M < \infty$ such that

$$\Sigma\, n^{-1}|f(I_n)| \le M$$

for every sequence $\{I_n\}$.

It should be noted that the intervals I_n are not assumed to be directionally ordered, i.e., indexed from left to right or from right to left. If we make this additional restriction in the above definition, we obtain the class OHBV [9], which clearly contains HBV. Belna [3] has shown that $OHBV_C \supsetneqq HBV_C$. Waterman [11] has shown that

$$OHBV_C \supsetneqq UGW \supset HBV_C$$

and

$$OHBV_C - GW \neq \emptyset .$$

It has been asked if the inclusion $GW \supset HBV$ is proper. That it is proper should have been seen from [2], where it was shown that

$$GW_C \supsetneqq UGW .$$

The argument there is, perhaps, overly abbreviated. Here we shall improve on this result and supply a detailed argument. We shall prove the following.

THEOREM 4. $GW_C - OHBV_C \neq \emptyset$.

That this is an improvement of the above is easily seen if we write it in the form

$$GW_C \supsetneqq GW_C \cap OHBV_C \supset UGW .$$

In this connection it is interesting to note that UGW can be characterized in terms of a variational property. Let

$$V_{OH}(f,I) = \sup\{\Sigma\, n^{-1}|f(I_n)|\}$$

where the supremum is extended overall directionally ordered sequences $\{I_n\}$ of nonoverlapping intervals in I . We have the following result.

THEOREM 5. $f \in UGW$ if and only if $f \in OHBV_C$ and $V_{OH}(f,(x,x+\epsilon))$ and $V_{OH}(f,(x-\epsilon,x))$ tend to zero with ϵ for every x .

PROOF. We have observed that $OHBV_C \supset UGW$. Consider only the behavior to the right of a point x . Analogous arguments apply for the left. Suppose first that $V_{OH}(f,(x,x+\epsilon))$ is bounded away from zero for arbitrarily small ϵ . Then there exists a right system at x, $\{I_{ni}\}$, such that either $\{\Sigma\, i^{-1}f(I_{ni})\}$ or $\{\Sigma(k_n+1-i)^{-1}f(I_{ni})\}$ does not converge to zero, implying $f \notin UGW$. Next note that for a right system at x ,

$$V_{OH}(f,(x,x+\varepsilon)) \geq \max\{\Sigma\, i^{-1} f(I_{ni}), \Sigma (k_n+1-i)^{-1} f(I_{ni})\}$$

for large n, so that if $V_{OH}(f,(x,x+\varepsilon)) \to 0$ as $\varepsilon \to 0$, f must satisfy the right hand portion of the UGW condition at x. ■

We now prove Theorem 4 by constructing an example of a continuous function which is in GW but not in OHBV and, therefore, in neither UGW nor HBV.

Let $\{a_n\}$ and $\{b_n\}$ be sequences in $(0,\pi)$ with $a_{n+1} < b_{n+1} < a_n < b_1$ and $a_n \searrow 0$. Let $\{\lambda_n\}$ be a decreasing positive sequence such that $\Sigma\, \lambda_n < \infty$. Choose a sequence of positive integers $\{k_n\}$ such that

(1) $$\lambda_n \sum_{i=1}^{k_n} [(i + \sum_{j<n-1} k_j)\log(1+i)]^{-1} \geq 1 .$$

Let $x_{ni} = a_n + (b_n - a_n)i/2k_n$, $0 \leq i \leq 2k_n$. For $n = 1,2,\ldots$ and $1 \leq i \leq 2k_n$ set

$$f(x_{ni}) = \lambda_n [\log(k_n - (i-3)/2)]^{-1} \quad \text{for odd } i$$

and

$$f(x_{ni}) = 0 \quad \text{for even } i .$$

Let f be the 2π-periodic function which is linear on each interval $[x_{ni}, x_{n,i+1}]$ and equals zero elsewhere in $[-\pi,\pi]$. Clearly f is continuous.

If $\{I_i\}$ denotes the sequence of intervals in $[-\pi,\pi]$ on which f is increasing, indexed from right to left, then for $N > \sum_{j=1}^{n} k_j$, (1) implies that

$$\sum_{i=1}^{N} i^{-1} f(I_i) > n .$$

Thus f is not in OHBV and, consequently, in neither UGW not HBV.

Note that for any $x \neq 0$ there is a $\delta > 0$ such that f is linear in $(x-\delta, x]$ and in $[x, x-\delta)$. Thus the GW condition is satisfied for any system at an $x \neq 0$. Since $f = 0$ on $[-\pi, 0]$, the GW condition is satisfied by any left system at $x = 0$. We shall now show that the GW condition is satisfied by any right system at $x = 0$ and, therefore, $f \in GW$.

Suppose now that $\{I_k\}$ is a finite set of nonoverlapping intervals in $(0, a_m)$ ordered from left to right. If $I_k = (\alpha,\beta)$ is contained in no $[a_n, b_n]$, $n > m$, and $f(I_k) \neq 0$, then, letting α' be the least number of form $x_{n,2i}$ in I_k and β' be the greatest, we can replace I_k in the set by one of the subintervals $[\alpha,\alpha']$,

$[\beta',\beta]$ without diminishing the sum

$$\Sigma\, k^{-1}|f(I_k)|\,. \tag{2}$$

Thus we may assume that each $I_k \subset [a_n,b_n]$ for an $n > m$.

If an I_k contains an interior point of form $X_{n,2i}$, then, by a similar argument, it is easily seen that we can replace I_k by a subinterval with no interior points of that form without diminishing the sum (2). Intervals I_k with interior points of form $X_{n,2i-1}$ may now be replaced by subintervals lacking such points by an analogous argument.

Combining these observations, we see that, without loss of generality, we may assume that each I_k is in some $[X_{ni},X_{n,i+1}]$, $n > m$.

Consider those $I_k \subset [a_n,b_n]$ for a particular n. Write $H_i = [X_{ni},X_{n,i+1}]$. If $I_k \subset H_i$ for $s \le k \le t$, then

$$\sum_{k=s}^{t} k^{-1}|f(I_k)| \le s^{-1}|f(H_i)|$$

and so, if $I_k \subset [a_n,b_n]$ for $\ell \le k \le \ell'$, there is an increasing sequence of integers $s_k \in [\ell,\ell']$ and an increasing sequence of integers $i_k \in [0,2k_n]$ such that

$$\sum_{k=\ell}^{\ell'} k^{-1}|f(I_k)| \le \Sigma\, s_k^{-1}|f(H_{i_k})|\,.$$

If H_{i_k} and $H_{i_{k+1}}$ have common endpoint $X_{n,2j-1}$, then $|f(H_{i_k})| = |f(H_{i_{k+1}})|$ and we replace $s_{k+1}^{-1}|f(H_{i_{k+1}})|$ by $s_k^{-1}|f(H_{i_k})|$ in the above sum.

Denoting the endpoints at which $f \ne 0$ of the intervals remaining in the right hand sum above by $X_{n,2i_j-1}$, $1 \le 2i_1-1 < \dots < 2i_{r_n}-1 \le 2k_n-1$, we have

$$\sum_{k=\ell}^{\ell'} k^{-1}|f(I_k)| < 2\sum_{j=1}^{r_n} j^{-1}|f(x_{n,2i_j-1})|$$

$$\le \sum_{i=1}^{r_n} i^{-1}|f(X_{n,2(k_n-r_n+i)-1})|$$

$$= 2\lambda_n \sum_{i=1}^{r_n} i^{-1}[\log(r_n+2-i)]^{-1}\,.$$

Finally, for the entire sum (2),

$$\Sigma\, k^{-1}|f(I_k)| < 2\sum_{n=m+1}^{\infty} \lambda_n \sum_{i=1}^{r_n} [i\log(r_n+2-i)]^{-1}.$$

Choose the integer q_n such that $\frac{1}{2}r_n \le q_n \le \frac{1}{2}(r_n+1)$. Then

$$\sum_{i=1}^{r_n} [i \log(r_n+2-i)]^{-1} = \sum_{i=1}^{q_n} + \sum_{i=q_n+1}^{r_n}$$

$$\le [\log(r_n+2-q_n)]^{-1} \sum_{i=1}^{q_n} i^{-1} + (\log 2)^{-1} \sum_{i=q_n+1}^{r_n} i^{-1}$$

$$\le [\log(q_n+1)]^{-1}(1+\log q_n) + (\log 2)^{-1}\log(r_n/q_n)$$

$$< 2 + 1 .$$

Thus

$$\Sigma\, k^{-1}|f(I_k)| < 6 \sum_{n=m+1}^{\infty} \lambda_n = o(1)$$

as $m \to \infty$, implying $f \in GQ$.

BIBLIOGRAPHY

1. A. Baernstein, "On the Fourier series of functions of bounded Φ-variation", Studia Math. 42 (1972), 91-94.

2. A. Baernstein and D. Waterman, "Functions whose Fourier series converge uniformly for every change of variable", Indiana Univ. Math. J. 22 (1972), 569-576.

3. C. L. Belna, "On ordered harmonic bounded variation", Proc. Amer. Math. Soc. 80 (1980), 441-444.

4. C. Goffman, "Everywhere convergence of Fourier series", Indiana Univ. Math. J. 20 (1970), 107-113.

5. C. Goffman and D. Waterman, "Functions whose Fourier series converge for every change of variable", Proc. Amer. Math. Soc. 19 (1968), 80-86.

6. C. Goffman and D. Waterman, "A characterization of the class of functions whose Fourier series converge for every change of variable", J. London Math. Soc. (2) 10 (1975), 69-74.

7. R. Salem, "Essais sur les séries trigonométriques", Actualités scientifiques et industrielles, no. 862 (Paris, 1940), 1-85.

8. D. Waterman, "On convergence of Fourier series of functions of generalized bounded variation", Studia Math. 44 (1972), 107-117.

9. D. Waterman, "Λ-bounded variation: recent results and unsolved problems", Real Anal. Exchange 4 (1978-79), 69-75.

10. D. Waterman, "Fourier series of functions of Λ-bounded variation," Proc. Amer. Math. Soc. 74 (1979), 113-123.

11. D. Waterman, "On the note of C. L. Belna", Proc. Amer. Math. Soc. 80 (1980), 445-447.

DEPARTMENT OF MATHEMATICS
SYRACUSE UNIVERSITY
SYRACUSE, NEW YORK 13210

Contemporary Mathematics
Volume 42, 1985

SCHAUDER BASES FOR $L^p[0,1]$ DERIVED FROM SUBSYSTEMS OF THE SCHAUDER SYSTEM

Robert E. Zink

1. If one performs the Gram-Schmidt maneuver on the system of Schauder functions, a standard Schauder basis for $C[0,1]$, the orthonormal system that one obtains proves to be itself a Schauder basis for $C[0,1]$ (Franklin, [2].) It was shown by Szlenk [4], however, that this happy occurrence is merely a fluke, since one may perturb a Schauder system Φ in such a way that the perturbed system Φ_π is again a Schauder basis for $C[0,1]$, while the Gram-Schmidt orthonormalization of the new system is not.

On the other hand, the Franklin system (the G-S orthonormalization of the Schauder system) is also a Schauder basis for each of the spaces $L^p[0,1]$, $p \geq 1$, so that it is conceivable that $GS\Phi_\pi$ could be a basis for some of the L^p spaces even though it fails to be a basis for $C[0,1]$. Ancient results show that one of four situations must obtain: $GS\Phi_\pi$ is a Schauder basis for precisely those spaces $L^p[0,1]$ with: (1) $p \in [1,+\infty]$, where $L^\infty[0,1]$ is to be interpreted as $C[0,1]$, (2) $p \in (1,+\infty)$, (3) $p \in [\frac{\alpha}{\alpha-1},\alpha]$, for some $\alpha \geq 2$, (4) $p \in (\frac{\alpha}{\alpha-1},\alpha)$, for some $\alpha > 2$. In [5] Veselov has shown that each of these possibilities can be realized by appropriately choosing the parameters in Szlenk's example.

The present discussion deals with a different sort of disturbance of a Schauder system, Φ. In this schema, one first deletes some of the elements of Φ and then applies the Gram-Schmidt process to the residual system, Φ_ρ. The orthonormal system $GS\Phi_\rho$ (a generalized Franklin system) certainly will not be a Schauder basis for $C[0,1]$, but, if the L^p-closure of Φ_ρ were all of $L^p[0,1]$, for some values of p, it is conceivable that $GS\Phi_\rho$ could constitute a Schauder basis for those spaces. Indeed, following a trail marked by Ciesielski [1] in his work on the Franklin functions, one finds that it is possible to delete a surprisingly extensive collection of elements from a Schauder system and still have a residual subsystem whose Gram-Schmidt orthonormalization is a Schauder basis for each space $L^p[0,1]$, $1 \leq p < +\infty$.

An amplified version of this article, in which are given missing details of the proofs of the results stated below, will be published elsewhere.

1980 Mathematics Subject Classification. 42C05.

2. The Schauder systems herein considered are the usual collections of spike functions associated with a sequence $\{\pi_n = \{t_{nk}: 0 \le k \le 2^n\}\}_{n=0}^{\infty}$, of subdivisions of $[0,1]$, satisfying: $t_{00} = 0$, $t_{01} = 1$; $t_{nk} < t_{n+1,2k+1} < t_{n,k+1}$, for all $n = 0,1,\dots,$ and for all $k = 0,\dots,2^n-1$; $t_{n+1,2k} = t_{nk}$, for all $n = 0,1,\dots,$ and for all $k = 0,\dots,2^n$; and $\lim_n\|\pi_n\| = 0$. The first two elements of the system are the constant function $\varphi_{00} = 1$ and the identity function φ_{01}. The remaining functions are defined in blocks of sizes 2^{n-1}, $n = 1,2,\dots$; $\varphi_{n,k-1}$, the kth element of the nth block, takes the value 1 at $t_{n,2k-1}$ and has for support the interval $(t_{n,2k-2}, t_{n,2k})$.

THEOREM 1 (Ciesielski). If Φ be a Schauder system, if $\{K_n\}_{n=1}^{\infty}$ be the sequence of Dirichlet kernels associated with the corresponding Franklin system, and if the sequence $\{S_n: L^1[0,1] \to C[0,1]\}_{n=1}^{\infty}$ be defined by $S_nf = \int_0^1 K_n(\cdot,t)f(t)dt$, then $\|S_n\|_p \le 3$, for every $p \in [1,+\infty]$.

Because the Schauder system is total in each space $L^p[0,1]$, $1 \le p \le +\infty$, it follows that the Franklin system is a Schauder basis for each of these spaces.

From a study of the almost-everywhere convergence of Schauder series, initiated by Goffman [3], a simple criterion for the L^p-totality of a set of Schauder functions follows easily.

LEMMA 2. Let $\Phi_\rho = \{\varphi_1,\varphi_2,\dots\}$ be an infinite subsystem of a Schauder system, and let $\{E_n\}_{n=1}^{\infty}$ be the sequence of supports of the elements of Φ_ρ. In order that Φ_ρ be total in each of the spaces $L^p[0,1], 1 \le p < +\infty$, it is both necessary and sufficient that

$$(*) \qquad \mu(\limsup_n E_n) = 1.$$

(Indeed, should (*) fail to hold, Φ_ρ would fail to be total in each of the L^p spaces.)

The proof of the lemma proceeds along the following lines. From Lemma 1 of [6], it follows, in particular, that each element φ of $\Phi\backslash\Phi_\rho$ is the a.e. pointwise limit of a sequence $\{f_n\}_{n=1}^{\infty}$ of finite linear combinations of elements of Φ_ρ, such that $0 \le f_n \le f_{n+1} \le \varphi$, for all n. Hence, an application of the Lebesgue theorem of dominated convergence suffices to show that the L^p-closure of Φ_ρ contains Φ. Because Φ is a Schauder basis for $C[0,1]$, one concludes that Φ_ρ is total in each space $L^p[0,1]$, $1 \le p < +\infty$.

THEOREM 3. Let Φ_ρ be a subsystem of a Schauder system Φ. If Φ_ρ satisfies both (*) and

(**) for every $\varphi \in \Phi_\rho$ and every $\psi \in \Phi$,
if $\operatorname{supp}\psi \subset \operatorname{supp}\varphi$, then $\psi \in \Phi_\rho$,

then, for every p in $[1,+\infty)$, $GS\Phi_\rho$ is a Schauder basis for $L^p[0,1]$.

The class of residual systems to which Theorem 3 may be applied is a rich one, including, for example, those Φ_ρ obtained by deleting from Φ an arbitrary initial segment as well as those systems obtained by excising from Φ each element the closure of whose support contains an arbitrary fixed point of $[0,1]$.

The demonstration of Theorem 3 makes use of (**) in order to ensure that the method developed by Ciesielski can be applied to show that the sequence $\{\|S_n\|_p\}_{n=1}^\infty$ is bounded. This fact together with the L^p-totality of Φ_ρ, guaranteed by (*), yield the desideratum.

While the condition (*) is certainly a necessary one for the conclusion of the theorem, the following result shows that (**) is far too restrictive.

THEOREM 4. If Φ be the (standard) Schauder system associated with the sequence of binary subdivisions of $[0,1]$, and if Φ_ρ be the residual system obtained from Φ by deleting any one of its elements, then $GS\Phi_\rho$ will be a Schauder basis for each space $L^p[0,1]$, $1 \le p < +\infty$.

3. The proof of Theorem 4 is based upon an extensive sequence of computations. It seems odd that, despite the trivial nature of the perturbation performed upon Φ, so much effort should be required for this demonstration. Odder still, the resolution of the slightly more general problem, in which one deletes an arbitrary finite number of functions, seems to be a task of a much higher degree of complexity. One senses that a different, more general, mode of thinking about the problem may be needed.

For example, a most helpful general principle would be the following: If Ψ and Φ are denumerable systems of continuous functions, if $\Psi \supset \Phi$, and if $GS\Phi$ is a Schauder basis for some space $L^p[0,1]$, then $GS\Psi$ is also a Schauder basis for that space. Theorem 4 would follow swiftly from this "principle," since, if φ were the deleted function, it would suffice to delete, in addition all of those Schauder functions whose supports are supersets of the support of φ and then to apply Theorem 3. Unfortunately, this would-be principle is false, as one can learn from a brief examination of the aforementioned Schauder bases introduced by Szlenk. Each of these systems Ψ is obtained from a Schauder system Φ by taking

$$\psi_{n,0} = \varphi_{n,0} + \varepsilon_n \varphi_{n,2^{n-1}-1}, \text{ for } n = 2,3,\ldots,$$

and

$$\psi_{n,k} = \varphi_{n,k}, \text{ in every other case.}$$

By appropriately choosing the sequence of underlying subdivisions $\{\pi_n\}_{n=0}^\infty$ and the sequence of perturbation constants $\{\varepsilon_n\}_{n=2}^\infty$, one can obtain a Szlenk system Ψ for which $GS\Psi$ is a Schauder basis only for $L^2[0,1]$. On the other hand, the system Ψ obtained by deleting from Ψ all of the perturbed fun tions, $\psi_{n,0}$, n = 2,3,..., coincides with a Schauder system Φ_ρ to which Th rem 3 may be applied . Thus, $\Psi \supset \Psi_\rho$, and $GS\Psi_\rho$ is a Schauder basis for space $L^p[0,1]$, $1 \leq p < +\infty$, while $GS\Psi$ is a basis only for $L^2[0,1]$. Gene principles, alas, seem to be in somewhat short supply. Nevertheless, there must be a better mode of attack on the problem. Surely it must be true that Theorem 4 can be extended by permitting any finite number of deletions to be made. Indeed, one believes that Theorem 3 should continue to hold if one we to delete the hypothesis (**).

One final bit of strangeness is, perhaps, worthy of mention. The zero se of the elements of a generalized Franklin system may have a rather large int section. Indeed, one may arrange matters so that this intersection is a pre assigned, nowhere dense, perfect null set; the Cantor set being a case in po To do this, one deals with a sequence of (Schauder) subdivisions of [0,1] termined by the exceptional set E . From the Schauder system associated therewith one deletes those elements whose supports are not contained in the complement of E . The result then follows from an application of Theorem 3.

BIBLIOGRAPHY

1. Z. Ciesielski, "Properties of the orthonormal Franklin system", Stud Math. 23(1963), 141-157.

2. Philip Franklin, "A set of continuous orthogonal functions", Math. A 100(1928), 522-529.

3. Casper Goffman, "Remark on a problem of Lusin", Acta Math. 111(1964) 63-72.

4. W. Szlenk, "Une remarque sur l'orthogonalisation des bases de Schaud dans l'espace C", Coll. Math. 15(1966), 297-301.

5. V. M. Veselov, "On properties of bases after their orthogonalization Izv. Akad. Nauk SSSR 34(1970), 1416-1436.

6. Robert E. Zink, "On a theorem of Goffman concerning Schauder series" Proc. Amer. Math. Soc. 21(1969), 523-529.

DEPARTMENT OF MATHEMATICS
PURDUE UNIVERSITY
WEST LAFAYETTE, INDIANA 47907

ABCDEFGHIJ–AMS–898765